A WOMAN AMONG WOLVES

A Woman Among Wolves

My Journey Through Forty Years of Wolf Recovery

Diane K. Boyd

Foreword by **DOUGLAS H. CHADWICK**

GREYSTONE BOOKS
Vancouver/Berkeley/London

Greystone Books Ltd.
greystonebooks.com

Cataloguing data available from Library and Archives Canada
ISBN 978-1-77840-113-8 (cloth)
ISBN 978-1-77840-114-5 (epub)

Editing by Jane Billinghurst
Copy editing by James Penco
Proofreading by Alison Strobel
Jacket design by Jessica Sullivan and Fiona Siu
Jacket photograph of Diane howling from the author's personal collection
Text design by Fiona Siu
Frontispiece photograph of Diane skiing through lodgepole pines by Paula A. White
Photograph on page 214 of Mike Fairchild, wolf 8910, and Diane by Carol Schmidt
Map by Leslie Lowe

The chapter "Sage" has been adapted with permission from "Sage's Story" in
Wild Wolves We Have Known, edited by Richard P. Thiel, Allison C. Thiel, and
Marianne Strozewski, published by the International Wolf Center in 2013.

Printed and bound in Canada on FSC® certified paper at Friesens. The FSC® label
means that materials used for the product have been responsibly sourced.

Greystone Books thanks the Canada Council for the Arts, the British Columbia Arts
Council, the Province of British Columbia through the Book Publishing Tax Credit,
and the Government of Canada for supporting our publishing activities.

Canada

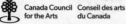

Greystone Books gratefully acknowledges the xʷməθkʷəy̓əm (Musqueam),
Sḵwx̱wú7mesh (Squamish), and səlilwətaɬ (Tsleil-Waututh) peoples on
whose land our Vancouver head office is located.

To Harold and Verna, because they didn't know;
and to the wolves, because they did.

CONTENTS

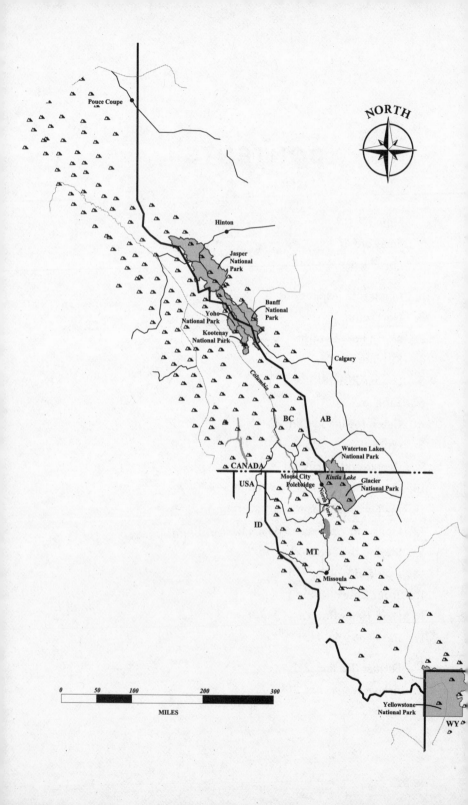

NORTH

Pouce Coupe

Hinton

Jasper
National
Park

Banff
National
Park

Yoho
National
Park

Kootenay
National
Park

Calgary

Columbia

BC   AB

Waterton Lakes
National Park

CANADA

USA

Moose City
Polebridge

Kintla Lake

Glacier
National
Park

North Fork

ID

MT

Missoula

Yellowstone
National
Park

WY

| 0 | 50 | 100 | 200 | 300 |

MILES

FOREWORD

F OR A GOOD span of years, I lived in a Montana cabin eighteen miles from Canada and just across the North Fork of the Flathead River from Glacier National Park. I was wandering the valley's bottomland forests in 1979 when I met Diane Boyd, a lone woman out searching for a lone female wolf named Kishinena. It was what she did nearly every day for three years running. Diane was well aware the animal she was tracking was no ordinary wolf. At the time, the only *Canis lupus* known to survive in the lower 48 states were in parts of northern Minnesota, where Diane first worked with this endangered mammal, and on an island in Lake Superior. Much of the public thinks the dramatic reappearance of wolves in the Mountain West after half a century was the result of a government program that released several dozen wolves from Canada into Yellowstone National Park and central Idaho during the mid-1990s. Yet wolves had reintroduced themselves naturally to Montana's North Fork more than a decade earlier and were already beginning to spread through the region. The wolves' return to a landscape their kind had once roamed was the legacy of long-solitary, silver-gray Kishinena, plus a black male who found his way to her side early in 1982, and their offspring, the Magic Pack. *A Woman Among Wolves* is the real, full story of the species' comeback—a detailed account of the colonists that survived to hunt and howl beneath moonlit peaks and of those that fell silent, mostly at the hands of people.

As Diane continues her North Fork studies for eighteen years, the book fills with other tales too, of a woman's trials both in the male-dominated wildlife management field and in tough physical settings. How do you approach a trap baited with meaty scents to catch a wolf for radio-collaring? VERY carefully. Are wolves that dangerous? Naw. But local grizzlies, lured to the same smells, can be. What about super-slippery solo river crossings in midwinter? Finally, she says:

> You reach the other side of the river and find a place on the shelf ice where you can carefully pull yourself up without it giving way. This is a difficult moment, when a misstep can dump you back into the river upside down wearing a full backpack. You crawl across the ice shelf, reach the shore, take off your chest waders, hang them in the willow bushes, and put on your ski boots and skis. "Well, that wasn't so bad," you think to yourself, because you are young and immortal, and you ski off to find wolf tracks.

Bushwhacking beside this scientist, you'll learn about the effects wolves can have on competing predators, scavengers, prey, and plant communities. You'll also find the new packs challenging human communities to rethink what saving America's wildlife heritage actually means. Where might we allow thriving populations of these wild canines to reclaim their keystone role within native ecosystems? Anywhere? How often will we choose to restrict them instead? How severely? In her straightforward, personable style, Diane explains the controversies without getting dumped upside down back into the river of myths, opinions, and political maneuverings. For extra perspective, she takes us with her to visit researchers in Romania and northern Italy, where people, livestock, and wolves have shared countryside for generations. Later in her career, Diane became a wolf and carnivore specialist for the state of Montana, only to find debates over the future of *Canis lupus* growing ever more heated. Many human beliefs and behaviors can seem terrifically hard to make sense of. But by the time you finish reading *A Woman Among Wolves*, you'll understand the remarkable animals themselves—more intimately and accurately than you might have imagined was possible. And this . . . this is the legacy of the lone woman I met decades ago out in a big, wild valley on her quest for tracks of the lone wolf called Kishinena.

—DOUGLAS H. CHADWICK, author of *Four Fifths a Grizzly*

INTRODUCTION

MY PICKUP BANGED and rattled along the potholed Inside Road in the northwest corner of Glacier National Park. Boxes of wolf traps and jars of bait slid across the truck bed. I was in a hurry, my mind focused on the wolf caught in a trap somewhere ahead in the lodgepole pine forest. Out of the corner of my eye, I noticed motion in my rearview mirror. I looked up to catch the glassy reflection of vivid yellow eyes framed by a wolf's black face looking over my shoulder from the back seat. How did I get here?

IT WAS A warm June evening, with the lingering pink light of a long, northern summer twilight. The nighthawks had started their evening hunts, swooping through the banquet of mosquitoes above the river. I had just finished checking my wolf trapline along fifteen miles of a rough dirt road and was about to fix a late supper. I was staying at the ranger station at the Polebridge Entrance, which leads into the undeveloped, primitive part of Glacier National Park. A car pulled up to me and an anxious woman jumped out, running toward me and shouting, "Somebody is illegally trapping wolves in the park. There's a gray wolf in a trap along the road. It needs help!" Thanking her for the report, I explained that I had set the trap for research purposes to learn more about the wolves' ecology and survival. I addressed her concerns and assured her that I had no intention of harming wolves. This calmed her down.

"How far up the road is the wolf from here?" I asked.

"I don't know. Quite a ways."

"Okay. Would you like to follow me to help fit the wolf with a radio collar and release it on-site?"

The woman regretfully replied that she would love to, but she had a long drive that night and had to keep traveling to make her connections. I thanked her for her time, apologized for any discomfort she felt, and assured her that I'd go find the wolf immediately and take care of the situation. The woman left, and I jumped into my truck, heading up the road as fast as conditions would allow.

While looking for the gray wolf, I checked all the traps that I had inspected just an hour before. On the left side of the road, in the fourth trap, was a coal-black wolf, hopping around with a front foot caught. It was going to be a long night. I readied the tranquilizing drugs, jab stick, capture kit, and radio collar and approached the wolf.

Knowing that I still had a gray wolf ahead, I decided to drug this handsome black wolf quite lightly so he would be able to wake up and walk away sooner. Down he went into sedated slumber. I measured his vital signs to make sure he was handling the sedative well, took my samples, fitted him with a radio collar, and stood back. He was limp—definitely not ready to get up and walk away. I couldn't leave him unattended and defenseless in the forest in case a bear, mountain lion, or another wolf came along and attacked him. And I still had the gray wolf to find, where I would repeat this whole scenario and hopefully finish before it grew dark. I had no time to waste.

I loaded the drugged wolf onto the back seat of my pickup. As I maneuvered his lean, athletic body into the truck, I noticed that he was beginning to jerk and make small movements with his head. I jumped into the front seat and started to drive, looking at my watch, figuring the wolf would be waking up soon. What I had not counted on was that the more the truck jolted and banged along, the more stimulated the wolf became, waking him early out of his narcotic stupor. I didn't realize this because I was concentrating on slaloming around potholes, looking for my traps, and watching out for oncoming cars on the narrow one-lane road.

The wolf sat up abruptly in the back seat, awake and alarmed to find himself in a Ford F-150 on the way to Kintla Lake with some distracted blond woman driving a bit roughly. When I saw his face suddenly in the rearview mirror, he was sitting up like a big German shepherd, looking over my right shoulder. I hit the brakes and swerved to the side of the road. I grabbed my catchpole from the back of the truck, cautiously opened the rear door, slipped the noose around the wolf's neck, and gently pulled. He resisted a bit but finally stumbled out onto the dirt road as I released the catchpole. He wove his way through the willows and was gone.

I jumped back into the truck and drove to the end of the road at Kintla Lake, checking all my traps as I went. No second wolf. I was relieved but puzzled. Then I realized that the woman had said "gray wolf" to describe the species and not the color of its fur; she had definitely seen this black "gray wolf." Communication! When it comes to wolves, it's all about communication in so many ways.

MY DAD ONCE told me that with me as his daughter, he always had something interesting to write in his Christmas cards. Likewise, my brothers, Jeff and Terry, and many friends have long told me I should write up my stories about the early years of wolf recovery in the North Fork. I kept copious field notes and journals and published many scientific papers in peer-reviewed journals. I loved what I did, and I loved talking to people about my unusual profession, but I didn't know if I had the wherewithal to write it all down. However, the story of the North Fork wolves is unknown to 99.9 percent of the world, and many of the early researchers who found and followed these wolves have since passed and cannot speak of our hope and our hardship. I feel it is up to me to tell how a few adventurous

Canadian wolves trotted south into the northwestern corner of Montana, filling the landscape with their howls and progeny fifteen years before wolves were reintroduced to Yellowstone.

THIS STORY IS about me, my friends, my mentors, and—most importantly—the special wolves whose paths I crossed. It is also about following my dreams no matter how difficult the path, about fighting against stereotypes while pursuing my passion for wolves, and about finding home—for the wolves and for myself.

The more I thought about sharing this story, however, the more I realized that it is much bigger than just the North Fork and Glacier National Park—the area of Montana where I got my start and blossomed as a young woman and professional biologist. In the 1970s, after the Endangered Species Act was passed in the U.S., wolves began recolonizing landscapes across North America and Europe from which they had been extirpated a hundred years earlier. Our wolf research in the North Fork was just the beginning of global wolf recovery. Now, at this point in my career, I want to open the minds of a wide variety of readers to consider the challenges involved when wolves appear on the landscape.

I want you to be there with me as I make icy river crossings, get my butt chewed out by angry hunters and ranchers, hear wolf howls reverberate around the mountains, ski along wolf tracks to find their feast of deer, and learn about the science and ecology of wolves. I want you to share my laughter, tears, frustrations, and pure joy. I hope these stories, spanning more than four decades of traveling the rocky road of wolf recovery, inspire conservation dreams and actions in the next generations, who will make critical decisions and craft solutions for future wildlife-human interactions. Finally, I hope that this story provides a nuanced understanding of wolf recovery that will make someone who is contemplating killing a wolf pause—and then decide not to pull the trigger.

1

MINNESOTA INSPIRATIONS

PUNGENT SMELLS FILLED my nostrils as I waded through sneaker-sucking cattails in the magical swamp only four houses away from my home. More than one of my Red Ball Jets had been lost in the ooze. Some of my finest childhood days were spent mucking around in decaying vegetation on the trail of water bugs, ducks, salamanders, butterflies, and frogs. When I was six years old, I took four warm eggs from a down-lined mallard's nest. I left several for the mama duck to raise in the wild, while I planned to raise my own cute, fuzzy ducklings at home. I pocketed my egg treasures and set them in a cozy nest I made in my bottom dresser drawer. Three days later, my mother discovered my nest and I had to throw away the cold, dead eggs before they exploded among my socks. When I learned that I had killed the baby ducks instead of saving them as I'd intended, I was crushed. It was an early lesson in biology that Mother Nature does better at creating ducks, and most other things, than a six-year-old human.

Caterpillars could turn into butterflies—how amazing! I collected dozens of caterpillars, along with the twigs and leaves they were on—putting them into my mom's quart-sized Mason jars. Dad helped me to poke holes for air in the metal sealing lids with a nail and hammer; then we twisted on the canning rings and set the jars on a shelf to see what would emerge. I don't think any of the caterpillars survived their translocation to my bedroom terrariums, but the possibility of hatching a butterfly enticed me to keep collecting them—just in case.

My parents and our neighbors had bought their suburban Richfield homes in a new subdivision just south of Minneapolis immediately after World War II. It was a great time to grow up, in the baby boomer era before the fears of "stranger danger" became a basic rule of American culture. My two older, red-headed brothers and I had fifty-four kids our ages to play with in the neighborhood block. We ran wild all summer through unfenced yards, playing army with air rifles that we pumped up, filled with dirt, and shot at each other (pre-OSHA). On warm summer evenings we played "Flashlight" and "Starlight, Moonlight, Hope to See a Ghost Tonight." My favorite high-stakes hide-and-seek game we called "The Fugitive," named after the popular television drama about the wrongfully convicted Dr. Richard Kimble, who escaped from death row and went on a wild chase of his wife's real killer every episode, all while evading the law. I was learning the ways predators capture prey, albeit through humans.

Mrs. Anderson and the Longleys were among my favorite adult neighbors. Mrs. Anderson would put the top down on her sporty, red 1966 Mustang convertible, load up her daughter Patty and me, and take us to nearby Bush Lake. She would sun on the beach while us girls went swimming and tried to catch sunfish in our hands. I never tired of trying to catch fish and giggled when they swam around my legs. The Longleys were a younger minister couple next door and had the most beautiful singing voices I'd ever heard; holy harmonies rolled out of their kitchen windows all summer. Mr. Longley was the only man I'd ever seen doing push-ups while in a handstand, with his legs straight and feet high overhead. The other neighborhood dads were all middle-class workers who put in long days on the job. Physical fitness was low on their list of priorities; they worked hard and spent summer evenings relaxing after work, sharing drinks, smokes, and laughter with neighbors on the patio.

My favorite summer pastime was going to the Buddays' cabin with my family and spending a week fishing from a motorboat that put-putted all over Lake Ida with a six-horsepower Evinrude motor. I was the chief frog catcher and digger of worms for fish bait. But the most fun was shooting Dad's .22 single-shot Winchester rifle at cans and bottles that we set up on a stump in the woods. I was nine years old the first time my dad handed the rifle to me and patiently showed me how to safely handle, load, and unload each shell. When he was sure I got it, he showed me how to look down the barrel through the open sights at the target and then pull the trigger. Large, rusty cans and shattered glass bottles fell off the stump.

My older brothers were much better shooters than me. I remember one shooting match between my dad and my brother Jeff (who has never killed a thing in his life)—they started out shooting cans, which got progressively smaller, until Dad stood a Heinz ketchup bottle cap on edge atop the stump. They both hit it. Then the ultimate challenge: my dad put a golf tee on its side, with the flat top facing the shooter. Jeff blew it into sawdust. My dad and I were both impressed.

I inherited my curiosity about nature from my dad, who grew up on a farm in Iowa. He loved to go pheasant hunting in the fall with my oldest brother, Terry, and the neighbors and their boys. I begged to go, but it wasn't a thing in those days for girls to go hunting with "the boys." I remember when the hunters arrived home with a trunkful of brilliantly plumed roosters that needed to be cleaned, and I pulled out those gorgeous, long tail feathers for my trophies. My dad showed me how you could pull the tendon high up in the rooster's long running legs and make the toes on the foot contract and release, contract and release. I ran around with a leg and showed off what I had learned about pheasant feet—big stuff for me.

My dad was forty-five when I was born, and by the time that I had completed my hunter's safety course and was ready to hunt, Dad was sixty years old and had given up bird hunting. But I made up for it as an adult, raising and training my own bird dogs and spending decades of my life in beautiful landscapes hunting pheasants, sharp-tailed grouses, gray partridges, and ducks with my precious German wirehaired pointers. Dad would have loved it, if he had lived long enough to join me.

RICHFIELD WAS HOME to a beautiful wetland and lake complex, Wood Lake, which was irresistible and only one block away from my home. After the I-35w freeway was built, cutting off the lake complex from our houses, us kids would run across the interstate, between the cars, to explore Wood Lake, which had adjoined my swamp before the freeway was built. Sometimes we would crawl through the street grates, enter the storm drains that ran underneath the freeway, and pop up on the other side like towheaded muskrats. It was glorious. I'm thankful that my parents never found out or I would have gotten a severe spanking.

When I was ten years old, I was playing in my backyard one day when I heard the rumble and clank of heavy machinery, and I ran down to my swamp. Men on huge machines were leveling and filling my special wild place. I was so bewildered and angry that I ran up to the man on a big tractor and yelled at him, above the roar of the machine, to STOP and leave my swamp alone. He idled down the engine for a couple of minutes and explained that it was his job, and he had to do this so that houses could be built here and make the place much nicer.

How could houses be nicer than wild, purple irises, or mothers with their ducklings creating a gentle wake as they swam along? Too soon my swamp was buried under sodded lawns, and four houses sat neatly on top of where I used to nestle among the wild cucumbers and tadpole ponds. I didn't understand why this had happened. I learned then that human developments and profits often trump wildlife and nature's beauty. At least the marsh complex on the other side of the freeway became the Wood Lake

Nature Center, which still serves the community as an educational, public wetland—with interpretive trails, floating boardwalks, and places of discovery. But it is not as wild as my swamp was.

AS A CHILD I was not allowed to have a dog, despite my desperate begging. When I was twelve, I bought my dad a Labrador-Irish setter mixed-breed puppy from Adams Doggie Shop for Father's Day. Of course, I really bought it for me. My dad took me to the pet shop the next day and made me return the puppy, despite my tears. When I was sixteen, my parents agreed it was finally time to get a family dog. A woman from my mom's bridge club had a scruffy dog that had just had a litter of six puppies, and we drove over and picked one out. We all fell in love with our black-and-white, sheltie-poodle crossbreed furball, who I named Duffy. I was careful with my precious cargo and took him to obedience school, where he graduated as the top dog in his class. I took him everywhere with me. He was smart and cute and became a twenty-five-pound family member whom we all adored.

When Duffy was three years old and I was nineteen, I took him over to Como Park Zoo in St. Paul, where they had just built Wolf Woods, a five-acre enclosure of forest and brush surrounded by a chain-link fence. This was the zoo's first venture into a more natural habitat, and it was a much better confinement than the despicable traditional display cages, where the other zoo animals lived on concrete slabs. I was fascinated by the wolf family in Wolf Woods and went to watch the three-member pack often: a typical gray, a black, and a creamy blond. I found the wolves mesmerizing, and I could watch them for hours. But sometimes they chose to hide in the dense foliage and wouldn't come out. I found that even more compelling, and I would hang around and wait for my wolfy friends to emerge.

The day I took Duffy to the zoo, we stood outside the enclosure. I had Duffy on a leash. When the wolves saw Duffy, they excitedly emerged from the woods and ran over to the fence. The wolves and the dog sniffed noses through the chain-link fence, Duffy dwarfed by his giant, yellow-eyed distant cousins. Duffy decided he had to scent-mark on the fence, practically in the wolves' faces. He cocked his hind leg and a stream of urine shot onto the fence by a wolf's snout. Quick as lightning, the wolf grabbed Duffy's tiny hind foot through the diamond-shaped opening in the fence and yanked my dog's leg through the opening into the pen. The other two wolves rushed over to join this live game of tug-of-war, no doubt the most fun they had had in a long time.

I dropped to my knees, cradling Duffy in my arms. I held on to his hind leg tightly where it had entered the fence, while Duffy screamed as the wolves tore his leg apart. There was lots of biting and blood everywhere. Two zoo visitors appeared out of nowhere, yelling, and threw themselves onto the fence, which gave in a bit— and the surprised wolves let go. I quickly thanked the two young men and rushed Duffy, now in shock, to the University of Minnesota's Veterinary Medical Center, just a few miles away. I felt disconnectedly calm as I rushed inside, carrying my injured dog.

I told the older vet what had happened, and his student vet quickly took Duffy into the hospital's surgery ward. As soon as they had whisked Duffy away to safety, I sat down and began to shake as my bottled-up feelings burst forth, overriding the adrenaline. The older vet sat down next to me and asked me when I had last had a tetanus shot. "Strange question," I thought. Then he told me to wash my hands in the restroom and come back out to him. I lathered up my hands in hot water in the clinic's bathroom sink and was totally surprised to see thirty-two puncture marks covering my hands. Now I understood his question about the tetanus shot, which I received the next day. In the trauma of the melee, I hadn't realized that Duffy had been biting me—lashing out in pain—as the wolves were mangling his leg.

The vets performed amazing emergency surgery for several hours on my little dog to repair his leg, which was broken in three places and required stitches deep into the soft tissue as well as the skin. When I picked Duffy up from the clinic the next day, he was dopey from the painkillers and wearing a cast. I asked the vet how many stitches they'd put in. "I quit counting after one hundred," he replied. The team did such a fabulous job repairing Duffy's damaged leg that a year later he didn't even limp, although the scars were visible for the rest of his life. I still have thirty-two scars on my hands to remind me of my stupidity during my first up-close-and-personal wolf encounter. I would have many more in my future, but none would be quite so harrowing.

IN 1970, MINNESOTA was the only state in the Lower 48 that harbored a viable wolf population, except for approximately three dozen wolves on Isle Royale in Michigan and a handful in Wisconsin. Despite my shaky beginnings with wolves, they continued to fascinate me, and I sought out opportunities to learn more about them when I was enrolled in pre-veterinary medicine at the University of Minnesota. I was fortunate that the world's foremost wolf expert, Dr. L. David Mech, had his office on the university's St. Paul campus, coincidentally only four blocks from the veterinary hospital that had saved Duffy's life. I stopped by Dr. Mech's office as a young, starry-eyed wolf enthusiast, of which he had undoubtedly seen too many. However, I wouldn't go away, and like a good parasite my modus operandi was to persist, persist, persist. Finally, Dave, as I eventually came to call him, came to see that I was an ambitious and determined young woman. He offered me an opportunity to volunteer on the Wolf Project, a captive wolf facility at Carlos Avery Wildlife Management Area, near Forest Lake, Minnesota, about a half-hour's drive from campus. The basic Wolf Project grew into a larger, better-equipped facility that became a nonprofit in 1990 known as the Wildlife Science Center that is still in operation today. One of Dave's graduate students, Jane Packard, was doing her PhD research there on two packs of captive wolves and several single wolves. She was looking at the relationship of hormones, status, reproductive suppression, and other things both fascinating and foreign to me. Yes, of course I would love to volunteer. How soon could I start? And so it was that I began my journey into the amazing world of wolf research.

DURING WINTER WOLF courtship and breeding season, a handful of keen college students and I waited as Dave and Jane darted and drugged wolves in their outdoor enclosures and then transported the tranquilized animals in a truck to a big, cold, concrete-floored shop where we students were excitedly waiting. The brilliant senior research scientist, Dr. Ulie Seal, and his equally sharp and acerbic wife, Marialice, led the blood sampling, injections, re-drugging, and precision test-tube labeling during our processing days. At very exact intervals, hormones were injected, blood samples were drawn, and correlations were later determined between the wolf's status in the pack, their hormonal responses, and so on. I was excited to finally be doing wolf research with top scientists, even if it was on captive animals.

I fondly remember those wintry days around wolf mating season when we all danced through an organized chaos of rapid-fire re-drugging, monitoring, and sampling of twenty tranquilized and waking wolves who were spread out on the floor a few feet apart. We drew blood, took rectal temperatures, inserted bladder catheters to collect urine samples, recorded heart and respiration rates, and then moved on to the next wolf, over the course of several hours.

We'd grab a bite to eat when we could amidst the wolf processing. Marialice would suddenly holler, "More drugs on 2307!" and I'd turn around to see that wolf 2307 was drunkenly crawling his way toward my unguarded backside. Ulie would rush over and inject a drug-loaded syringe into the wolf's rump as I moved away, and the wolf would slither back down. Amidst this bedlam, Marialice would shout, "Blood draw on 1449!" and one of us would grab a syringe and rush over to wolf 1449, drawing blood from the cephalic vein on a foreleg. At the time, I didn't think anything about pulling a rectal thermometer out of a wolf's butt, shouting out the time and temperature, grabbing a bite from a sandwich, and moving to the next drugged wolf. I loved it all. And it was definitely worth the less-than-stellar grade I received in the veterinary parasitology class that fell in the same time slot as my wolf processing days. I had my priorities.

I had started my pre-veterinary medicine curriculum in 1973, but after those Wolf Project processing days began in 1976, I switched to an undergraduate degree in wildlife management instead. I wanted to work with wildlife, not pets or farm animals. At this time, apart from those small populations in three states, there were no gray wolves anywhere in the Lower 48. I was fascinated by these wild, unseen predators of the north and wanted to continue my experience with them, but how?

AFTER I HAD diligently volunteered on the Wolf Project for a year and hadn't objected to smelling like wolf anal glands, urine, and poop, Dave knew he had a budding wolf biologist on his hands. He offered me a volunteer position as a technician at the Kawishiwi Field Laboratory (K-Lab) on his decades-long wolf study in northern Minnesota. This extremely wild corner of Minnesota hosted a robust wolf population, along with white-tailed deer, fishers, foxes, lynx, moose, and black bears. It was the chance of a lifetime that would open many doors for me along my career path. The position was unpaid, so I sold my beloved, orange Yamaha RD350 motorcycle to pay for my summer internship. I loved my set of wheels, but selling it probably saved my life. Thank you, Dave!

I borrowed one of my parents' cars for the summer and headed up to K-Lab, a remote U.S. Forest Service outpost just south of Ely, built in the 1930s by the Civilian Conservation Corps (CCC). It was quite late. I was driving the last few dark and twisting miles of

Highway 1 when a wolf darted out of the dense jack pine forest and raced across the blacktop into the beam of my headlights. I stomped on the brakes and just missed hitting the wolf with the front bumper of the car. The adrenaline rush gave me a giddy nausea, so I pulled over onto the shoulder to recover for a few minutes. I had just experienced my first encounter with a wild wolf in the boreal forest—and I had nearly killed it. But it was damned exciting.

The rest of the summer at K-Lab I worked with Dave's wolf technician Jeff Renneberg, as well as researchers studying black bears, deer, ravens, and other fascinating wildlife species. I learned to navigate the woods with maps and compass, long before GPS, and became confident trusting these simple tools while working my way through the deep forest. I learned how to use the lab's delicate telemetry antenna and receiver to dial in a frequency and listen for the steady beeps of a particular wolf. I watched as Jeff set dozens of wolf traps, carefully burying and disguising each trap and adding a few drops of his special wolf lures. After a month of patient watching, I worked up the courage to ask if I could please set a trap, so he reluctantly handed me a heavy, evil-looking Newhouse No. 14 leghold trap. I finally had my moment—only to find that no matter how much I struggled, I could not physically open the trap. He took it from me (annoyed), opened it, and laid another well-disguised wolf trap without any help from me. I was humiliated.

That evening when we got back to K-Lab, I walked down to my cabin with a trap concealed in my backpack. I pulled out that trap, then stood on it, grunted, pulled at the stubborn jaws, jumped up and down on it with all of my 130 pounds, twisted, cursed, and repeated the sequence many times. After half an hour of struggling, I finally leveraged the trap open by pulling up on the jaws with my hands while pushing down on the double side springs with my feet—and with those opposing forces in play, the jaws magically opened.

I quickly hooked the dog (lever) under the pan notch (trip mechanism), and the trap was set open. I was sweaty and proud. I reached underneath the jaws, tripped the trap trigger, and watched the jaws slam shut harmlessly. I wrapped the chain around the trap, turned to take it back up to the truck, and looked up to see Jeff leaning against a huge, old white pine tree with a long grass stem sticking out of his mouth, watching me and smiling ever so slightly. I stopped in mid-stride. He nodded silently to me, then walked up the hill to his truck and drove back to his home in Ely.

The next morning Jeff asked me if I wanted to set the first trap, so I did—with his clear approval. Opening traps was much easier once I figured out the secret: it's technique, not strength. After that, we worked as partners and he warmed up to me. I was on cloud nine until I finally caught my first predator: a skunk. Jeff was not pleased and muttered under his breath that he had never caught a skunk in a wolf trap. He picked up a rock, and with one throw he killed that skunk dead. I was sorry for the skunk but also totally impressed with Jeff's pitching arm. Unfortunately, the skunk had the final word as it sprayed copiously in its death throes. We got thoroughly doused in skunk essence as we loaded our fouled equipment back into the truck, finished checking the traps, and drove into Ely for breakfast. When we walked into the café, people looked up and sniffed—and those in the know deduced what had happened and snickered. Jeff just shook his head. Eventually, I caught a wolf and I redeemed myself, with some good-natured ribbing from Jeff.

MY PROFESSIONAL CAREER began at K-Lab running wolf traplines, hiking mossy spruce forests, howling and waiting for wolf pups to reply to confirm reproduction, collaring wolves and flying telemetry flights to listen for pings from their radio collars, necropsying snowshoe hares, eating communal dinners at the main lodge, and learning with a group of fun and interesting researchers.

Rustic K-Lab, with its mouse-infested accommodations, charmed us all and gave us amazing research opportunities among the pines and along the Kawishiwi River. This old CCC work project holds a special place in the hearts of scores of us who got our start on Dave Mech's Superior National Forest wolf study. The K-Lab closed in 2010. The buildings were scheduled for demolition due to their deteriorating condition, but researchers and supporters of the historic compound spoke up loudly; K-Lab was saved, restored, and added to the National Register of Historic Places.

I visited K-Lab in 2018 with Dave and some other researchers. The old log lodge was looking good, but it was locked up tight. We peered in the windows and shared our memories and stories as we walked among the old buildings. I looked up at that tree that Jeff had leaned against forty-five years earlier; the ancient tree was still there and taller, although Jeff had passed away many years before. Thanks Dave, Jeff, and Jane for giving me my first breaks as a wolf biologist.

AS PART OF our undergraduate wildlife curriculum at the University of Minnesota, my classmates and I spent six weeks at the Lake Itasca Forestry and Biological Station within Itasca State Park, about 220 miles north of Minneapolis. This remarkable and remote field station is located by the headwaters of the Mississippi River, near the junction of the boreal and hardwood forests and the prairies, a rare convergence of three very different ecosystems. Fabulous Itasca and its inspiring field-oriented professors served as the foundation for many of us future biologists. I fondly remember Dr. David Parmelee, a soft-spoken and superb ornithologist, and Dr. John Tester, a friendly and exuberant ecologist, who mentored us through delightful and difficult fieldwork.

Dr. Parmelee took us to all three ecosystems and showed us the birds in their native habitats. We were challenged to identify birds by their flight patterns and habitat as we zipped by in a van at fifty miles an hour. We studied nests, eggs, song recordings, and skins from the collection at the field station. We were assigned four-acre forested plots and were responsible for identifying all the birds within them. Not only that, but we had to show Dr. Parmelee every resident bird species there and their nests. He never missed a bird, but I did. Dr. Tester had us run mini-traplines with snap traps and live-capture Sherman traps to catch a sample of small rodents in the different habitats. I learned to quickly identify voles, deer mice, and shrews, and the larger live traps occasionally rewarded us with a chipmunk or jumping mouse.

My time in Itasca instilled in me a sense of awe and curiosity for different habitats. I learned about bird migration, mammals, trees, plants, fungi, lake ecology, hibernation, bats, tree identification, and just about everything else all around me.

I AM A strong believer in the power of education and have taught wolf and wildlife conservation classes throughout my career. I'm still teaching for the University of Montana's Flathead Lake Biological Station, the Rocky Mountain equivalent of the Itasca Biological Station. Students light up when they find scats, wolf-pup-chewed bones, and beaver skulls in an abandoned wolf rendezvous site— the equivalent of a wolf nursery, where the adults leave the pups behind with a babysitting adult while the pack goes off hunting. We debate the findings of professional papers as we discuss real and perceived negative impacts caused by the return of wolves. Where is the truth? How do we find solutions to wolf-human conflicts? How will they shape the world? Their wheels begin to turn, and I love being a part of it.

Fresh out of my first field courses at Itasca, I was ready to become a wildlife researcher and save the wolves, armed with a few basics and a hell of a lot of enthusiasm. Observing, catching, radio-collaring, and following wolves was going to be so exciting!

2

...

NORTHOME

I T WAS THE summer of 1979, and I sang along to Joni Mitchell and Gordon Lightfoot cassettes while driving the five hours north from Minneapolis to Northome, Minnesota. At twenty-four, I was freshly graduated from the University of Minnesota with my bachelor of science degree in wildlife management, and I was thrilled at the prospect of starting my first professional job: wolf technician in the rural, remote community of Northome.

Wolves were my passion and this was my dream job. I would be working for the U.S. Fish and Wildlife Service trapping and removing livestock-killing wolves, helping set up preventive measures to protect livestock, and trapping and radio-collaring other wolves for a research project. I would be working with wild wolves in the north country, on my own, and I was eager to tackle whatever the job threw my way. To the best of my knowledge, there were no other women trapping wolves in North America for livestock problems or research at the time. I had set my bar high, and I was determined to sail over it and land safely on the other side.

My new home was a one-room log cabin on an abandoned farm homestead eight miles outside of Northome, a conservative town with a population of 312. U.S. Route 71 and Minnesota State Highways 1 and 46 met in Northome, with Bartlett Lake bordering the east side of town. An uninhabited area of wetland and swamp known as the Big Bog began just north of town and stretched up nearly seventy miles to the Canadian border. The surrounding land had once been old-growth forest, but by the 1930s the massive red and white pines of the north country had been heavily logged. The timber industry had subsequently suffered, and people had turned to raising crops, cattle, and sheep—chopping pastures out of dense, second-growth forests of birch, aspen, spruce, maple, and fir.

When I arrived, it was a quiet community of farmers, loggers, resorters, and small business owners. Locals and tourists alike enjoyed hunting the forests, fishing the lakes, and relaxing with a cold drink at the local tavern. Other activities included attending the Koochiching County Fair, four-wheeling, or reading the weekly newspaper, the *Northome Record*, which reported on visiting relatives, local bridge club news, ice fishing contest winners, and café specials. It didn't take long to explore the one square mile of town, which comprised the hardware store, K–12 school, gas station, library, bank, Lutheran church, grocery store, beauty salon, and abandoned Burlington Northern railroad depot. The community took pride in its big bruins and called the area the black bear capital of the world. Watching enormous black bears feeding at the dump was standard family entertainment. The nearest stoplight was in Bemidji, forty miles away.

BEFORE LOGGING, the old-growth forests surrounding Northome had been home to woodland caribou, which were the predominant ungulate species in much of the current moose range in northern Minnesota. The caribou vanished by the 1930s after the removal of the mature conifer forests they depended on, compounded by unregulated hunting, climate change, and a hair-thin parasite carried by white-tailed deer. *Parelaphostrongylus tenuis*, or brainworm, is carried by white-tailed deer, which have developed a strong resistance to the parasite and remain basically unaffected by its presence. However, brainworm is quite lethal to moose and caribou. In an unanticipated turnover after the mature pines were logged off, caribou numbers dropped and white-tailed deer and moose flourished in the regenerating habitat of birch and aspen. The interspecies transmission of brainworm was likely the coup de grace for the caribou, and it eventually hit moose populations as well.

ANCIENT CARIBOU TRAILS were still visible in the Red Lake area near Northome, which served as a reminder of this vanished species. The evolving woods were now home to white-tailed deer, moose, foxes, and wolves. The rich, black, loamy soils were fertile ground for raising alfalfa, clover, and livestock. Cattle grazed unattended in the thick forests where the wolves denned and raised their pups. And that's how wolves got into trouble—it was hard for them to resist the slow groceries walking by.

Throughout North America, wolves have historically been routinely shot, trapped, hunted, and poisoned—until they became protected as "endangered" in the U.S. in 1974 under the Endangered Species Act. By this time, however, wolves had been extirpated from 98 percent of their historic range in the Lower 48, until fewer than a thousand of them remained in the more remote areas of northern Minnesota.

Folks in Northome hadn't yet embraced the concept of endangered species when I appeared on the scene. Legally, farmers couldn't harm wolves that were killing their cattle; they were supposed to call in the state and federal authorities to help them resolve depredation issues while the wolves chewed on their livestock. That created a lot of delay and ugly frustration, so "shoot, shovel, and shut up" was often the norm for dealing with problem wolves.

I worked with the local game warden, Lonnie Schiefert, who wasn't thrilled about working with a tall, blond college girl from the Cities (as the Twin Cities metropolitan area was disdainfully called) to help him resolve wolf problems. I think he was expecting a coarse, bearded, beaver-slinging trapper type. What a disappointment I must have been.

When a farmer telephoned Lonnie with a depredation complaint, Lonnie and I would rendezvous, meet the angry farmer, and look at the bloated, bruised, maggot-infested livestock carcass. Then we would skin the reeking animal to see if we could find identifying bite marks and hemorrhages. We would look for wolf tracks and blood trails in the vicinity, and together we would determine cause of death. Pretty exciting stuff for a budding wolf biologist raised in suburban Minneapolis. If we determined that the animal had been killed by wolves, I would set traps within the boundaries of the farm for up to ten days and try to catch and remove the offending wolf (or wolves) so it (or they) could be euthanized—while the non-depredating wolves could remain unharmed, hopefully without developing a taste for livestock.

These were the guidelines mandated by the Endangered Species Act, guidelines we were all supposed to follow. Farmers were compensated for wolf-killed livestock, and the payments were intended to help offset the trauma. But some farmers I talked with told me they raised their cows to be eaten by people, not wolves, and they didn't want the damn compensation. If Lonnie and I determined a cow or sheep died of causes unrelated to wolves, I didn't set traps and farmers were not compensated. Of course, the farmers assumed all dead livestock died at the fangs of wolves, and our determinations to the contrary did not sit well.

IT HAD SOUNDED straightforward enough in the job description, but by my second week on the job, Lonnie looked me straight in the eye and told me that he didn't believe that I could catch a wolf. Say what? I had mistakenly thought he was going to be my colleague, not my adversary. His doubts did nothing to boost my self-confidence in what was already a challenging position.

In truth, I had learned to trap wolves only two summers before as a volunteer on Dave Mech's wolf research project 125 miles to the northeast, and I had only caught a handful of these cunning creatures. I indignantly told Lonnie that I had indeed trapped lots of wolves.

Lonnie sneered, "I'll bet you five bucks that you don't catch one here all summer."

My hackles were up. I said, "You're on."

As luck would have it, three days later I trapped my first Northome cow-killing wolf, so I pulled a hair sample and took Polaroid photos of the unlucky animal. I drove to Lonnie's house, gave him my wolfish evidence, and waited. He shook his head, pulled out his billfold, and handed me a $5 bill. We got along much better after that. Word got around Northome that I had begun to catch wolves, and people sometimes began to acknowledge me at the post office and café, asking how things were going. I was grateful for the slight positive shift in local attitudes.

I'd been working in Northome for a month when an older fellow at the gas station told me that I "oughta buy a copy of the local paper and see what Bing wrote about you." Bing Elhard had been born in Northome six decades earlier. This town was his home and his weekly column in the *Northome Record* was gospel. I bought a paper and in it there was a large photo of this old guy, Bing, standing next to a six-foot-tall cardboard cutout of a grinning wolf, which reminded me of Wile E. Coyote. I read Bing's words about "the new attractive blond lady wolf trapper" who had come to Northome, and my face turned crimson. Bing stated that he had a serious wolf problem in his yard and challenged "the lady wolf trapper to come pay him a visit and solve the problem." I felt a knot grow in my gut. How I dealt with this could make or break my success here. What to do?

I spent that evening in my cabin building a custom wolf trap for Bing. I paid him a visit the next day. As I pulled up to Bing's house in my federal truck, he stepped out onto his porch with a smirk, arms crossed, and hollered, "You must be the lady wolf trapper!" I said, "Yes I am. Pleased to meet you." I circled around to the back of my pickup and pulled out my nearly perfect cardboard replica of a Newhouse No. 14 wolf trap, wrapped in black electrical tape to look like a black-dyed steel trap, and a Mason jar full of nasty goo. I walked up to Bing with my trap and bait and stated, "This is the best trap to catch the particular subspecies of wolf that has been bothering you, and this jar contains foul-smelling cardboard bait that I guarantee will lure your wolf into the trap."

I held the trap and bait out to him as we stood there, five feet apart. His jaw dropped. He stared at me incredulously for a long while, and then he broke into a grin, put his hand on my shoulder, and said, "C'mon in. The wife just pulled a blueberry pie out of the oven." We sat in his kitchen and got acquainted over coffee—and some of the best pie I have ever tasted. To Bing's credit, in his column the next week he published a photo of my cardboard trap and said his wolf problem had been solved. I could have hugged him for giving me that critical break into the Northome community.

WHEN I WASN'T searching for wolves and investigating livestock depredations, I immersed myself in the world of the wild things all around me at the cabin. I spied on a fat, glossy black bear walking through the aspens next to the cabin, rooting through the vegetation with its sensitive tan snout caked in moist soil. I drifted off nightly to the breezy rustle of the aspen leaves and the occasional great horned owl hooting across the clearing.

One evening at dusk, I was walking my big, shaggy mongrel dog, Stony, when I discovered a male woodcock claiming a small clearing in the woodland as his territorial mating ground. The bird slowly circled upward until almost out of sight, before beginning his display flight—a crazy, high-speed, downward spiral. He repeated this amazing exhibition over and over as he twittered madly and plunged into the grass. As the amorous woodcock began his eccentric ascent, I began to anticipate where he would land. Then I dashed over and hid in the grass so I could watch him plummet to earth a few feet away from me. My heart raced as the woodcock resumed his *peent* call from the leaf litter, beginning the courtship show anew, oblivious to my presence.

Other memories were less pleasant. I checked my traps alone, and one day while checking my trapline on a deeply rutted road in the back forty, a farmer followed me. I stopped and sat in my idling truck while he got out of his battered pickup and slowly walked over to me. He was a really big man and I sensed trouble on his face. I rolled my window all the way down to chat and did my best to be friendly to mask my apprehension. His body posture scared me as he leaned in on my lowered window. He began talking slowly about the weather and wolves, and then suddenly grabbed my thigh with his enormous, strong hand. I punched the accelerator hard and nearly broke his arm on the doorjamb as I sped away, mud clods flying, my heart pounding with terror. I managed to complete my trapping stint there in a week while avoiding him. I was too scared to tell anyone about the incident, worried that if word got out it would spawn copycat incidents for sport, or that my boss would feel that hiring a woman was a mistake. I dreaded both equally, and I felt vulnerable and alone.

Other livestock producers were easier to work with and supported me while I tried out some non-lethal aversive conditioning techniques to dissuade wolves from attacking livestock. These efforts included wrapping tennis-ball-sized wads of raw hamburger laced with lithium chloride in fresh pieces of cow hide, tying the beefy packets with string, and putting them out where wolves would discover them. The theory was that the lithium chloride would make wolves violently ill shortly after consuming the bait, and they would associate the taste of beef with ferocious vomiting— hopefully teaching them that cows were bad business and decreasing their desire to bite one. I put out hundreds of these bait packages and all were quickly scarfed up—with no decrease in depredations. Nice try, but the baits hadn't worked as planned. I also put out blinking highway lights around pastures that wolves had been visiting, as well as flapping rope lines of flagging known as fladry, and noisemaking devices.

One cooperative farmer was happy to allow me to try everything in my toolbox to help keep his cows safe. A few days after setting up a dozen blinking, orange highway lights along the edges of his long, skinny pasture, I stopped by to visit with him. I asked him if he'd seen any wolves lately and he said, "No." I was pleased with this report and asked him brightly if he thought my aversive conditioning efforts were working. He replied with a twinkle in his eye, "Don't know about that but I darn near had a plane land in my pasture last night." We both laughed as this old farmer tacitly acknowledged that I had a tough job to do and it was okay to have a little fun doing it.

AS I WAS still finding my way within the Northome community and figuring out my work routines, my supervisor at the U.S. Fish and Wildlife Service, Steve Fritts, had patient faith in me and allowed me a huge amount of independence. Sometimes there weren't enough depredation complaints to keep me busy, so I was overjoyed when Steve said that he wanted me to trap and radio-collar non-depredating wolves for his research project. Fabulous! Now I was on the road to becoming a real wolf researcher. Catching and collaring wolves to learn about their lives was much more appealing to me than catching problem wolves to be euthanized.

I explored rough gravel roads in my four-wheel-drive pickup, searching for fresh wolf scat and urination scent posts. I howled in remote areas in the hopes of getting a pack of wolves to howl back, tipping me off to where to set my traps. One day, I howled from a potholed road cutting through a cedar swamp. I waited a few minutes and soon heard brush crashing and sticks breaking to my right, followed by a wolf pup galloping out onto the road, looking for his family member who was bringing home a deer leg or a fatty beaver. With a happy face and lolling tongue, the gangly pup looked around for lunch. The pup spotted me, ears pricking forward, mouth closing, yellow eyes staring; then it spun around, laid its ears flat, and sped back into the cedar swamp. I knew that I had found my first wolf rendezvous site. I quickly set three traps along the shoulder of the road in the hopes of capturing an adult, and then I left the area to return home.

I slept fitfully that night with anticipation of catching a hand-some, eighty-pound adult wolf. I got an early start the next morning and drove to the cedar swamp. I had indeed captured a wolf in my first trap, but it was a pup. As I stood there admiring her wild but undeniable cuteness, I saw her big brother jumping around in my second trap, set seventy-five yards down the road. I didn't want to drug the first pup because I wasn't sure if I should radio-collar her or not. So instead, I secured her with rope hobbles around all four legs and a quick, loose wrap of electrical tape around her muz-zle, which I could easily remove with minimal pulling of her snout hair. I loaded her onto the front seat of my federal truck, and she submitted completely to me, no doubt terrified of the first human she had ever seen, let alone the first to give her a ride in a pickup. I drove up to the second pup and was getting ready to secure him when I noticed movement down the road—there was a third pup in my last trap. It was turning into a busy day. I secured the second pup, undrugged, with rope and electrical tape, loaded him onto the seat next to his sister, and drove up to the third pup. I taped and tied him and, since the front seat was full of wolf pups, I placed him on the floor on the passenger side.

The three pups weighed twenty-nine to thirty pounds each, which put me in a quandary. Steve had told me that if I caught any wolf pups I shouldn't radio-collar them unless they weighed at least thirty pounds, so that they would be large enough to wear a padded radio collar without being too encumbered by it. A growing pup would need to have a radio collar fitted to an adult circumference so it wouldn't choke, but one that also temporarily fit snugly enough to not come off over its head. This was accomplished by setting the collar to an adult-size circumference, then filling in the inside of the collar with foam padding and securing it with duct tape wrapped around the foam and collar. Hopefully the duct tape and foam would rot off by the time the pup was adult-sized and not before, or else the collar would slip over the pup's head prematurely and fall off.

Unsure if I should collar all three pups or none of them, I drove to the nearest gas station, left the trussed-up pups resting on the seat and floor, and went inside to call Steve on a pay phone and see what he wanted me to do. We had been talking for about ten minutes when I looked outside and noticed a crowd gathered around my pickup. Between people's shoulders I could see bouncing movements inside my truck.

"Uh, I gotta run now, Steve. Will call you later. Bye!"

I rushed over to my truck to see that the pups had wriggled free of their bonds and were wildly bouncing around the front seat, enthusiastically biting anything they found. I carefully opened the door and wedged a large, aluminum-framed backpack between me and the pups; they cowered away from me on the passenger side. I drove a couple of miles down the road, pulled over, loaded a syringe with a tranquilizer, and fastened the syringe to the end of a three-foot pole to poke each pup as it orbited around the cab. After fifteen minutes of wolf wrangling, I had all three pups drugged and draped across the seat.

I radio-collared all three, with the foam padding to allow room for growth, then drove the short distance back to where I had captured them and let them out. I watched them quietly from inside the truck, at a distance of seventy-five yards, wanting to make sure they woke up safely in their own time. After an hour, they wobbled up and drunkenly walked off into the thick forest, headed for home. I never told Steve what had happened inside the truck. Years later, rumors came back to me about how my dog had chewed up the steering wheel and the stick shift in the federal work truck that summer. I never corrected this misperception because my own stupidity seemed worse to me than the dog-bite story. Let sleeping wolves lie.

AS SUMMER WORE ON, I slowly became a bit more accepted in the community as I earned some credit for my trapping skills. I captured several livestock-killing wolves, which I removed to be euthanized, as well as wolves that I collared and released into the national forest for research. But I didn't tell anyone when I incidentally captured a nursing female coyote who had the misfortune to be feeding on a cow killed by wolves. I quickly released her so she could run back to her pups with a bellyful of beef and teats full of milk. If the farmer had known what I'd done, he would not have been happy. To him, coyotes were just the cowering cousins of the wolf devils. But this coyote was simply in the wrong place at the wrong time, and I saw no justification for harming her.

After I had been working in the area for a couple of months and was feeling more comfortable, I stopped in at the local watering hole to play pool and enjoy some socializing. After my first pool game, I noticed that there were five tequila sunrises lined up on the edge of the pool table. I'm not much of a drinker but I didn't want to offend folks during my bar debut. I slowly consumed one drink and gave the other four drinks to nearby patrons. I may not have impressed them with my drinking prowess, but I was a good pool player and beat some of the locals, which gave rise to some good-natured kidding. Out on the road, a few farmers in their pickups started to raise a hand from their steering wheels, giving a hello wave as we passed. Even Bing followed up and wrote a couple of neutral columns about my work, reporting my capture of wolves on his neighbors' farms. This was better than I had expected.

In August, I received the letter I was hoping for from the University of Montana's wildlife biology program, congratulating me on being accepted into their master's degree program. Hallelujah! I was to study the first recolonizing wolf in the western U.S., who had just set up a territory along the border between Montana and British Columbia, Canada. I finished up my Northome projects in early September, packed up my dog and belongings, and hit the road to begin my wolf research in the Rocky Mountains near Glacier National Park. I don't know if stories persisted in Northome about the blond lady wolf trapper, stories told with some head-shaking and maybe even a little acceptance—at least that's how I saw it in my dreams. But for me, that Northome summer was a crucial one for proving myself. I found professional validation in developing creative problem-solving skills, building bridges in the community, and learning about the complex relationship between wolves and humans. I had to constantly reconcile my love and admiration for these stunning animals with the removal and killing of problem wolves, so that their non-livestock-killing brethren could live. It was all part of my baptism by fire in the wolf politics of rural North America.

If someone had asked me in 1979 if I thought wolves would spread throughout the Midwest and fill the mountains and river valleys of seven western states, I would have laughed out loud. But these resilient and clever animals were tougher than I had realized. I also had no inkling that my enthusiasm to understand wolf recovery would become my life's work. And it all started in a place I'd never heard of called the North Fork of the Flathead River, along the northwestern boundary of Glacier National Park.

3
...

NORTH FORK
BAPTISM

B Y THE 1930S, the last viable population of wolves had been
extirpated in Montana and the rest of the Lower 48—except
for several hundred that survived in northern Minnesota,
and a few packs that persisted in northern Wisconsin and Michigan's
Upper Peninsula until the 1950s. The cruel campaigns of trapping,
poisoning, and shooting took their toll until the plains, mountains,
and even the national parks were wolfless. It is a sad saga chroni-
cled in many history books, with gruesome photos of dead wolves
and proud men. A few stragglers continued to find their way south
from the bountiful wolf populations in Canada, but most of those
Canadian wanderers didn't survive long once they crossed south
over the forty-ninth parallel. The last of the loners were loathed for
their livestock-killing, while simultaneously revered for their wild
cunning. The legendary White Wolf of the Judith Basin, killed in
May 1930, is mounted, with a permanent snarl, inside a glass display
case in Stanford, Montana, at the Basin Trading Post, where he is
still the small town's leading attraction.

But social perspectives changed, and as of 1974, the Endangered Species Act protected gray wolves throughout the Lower 48.
Bob Ream, a University of Montana professor in the school of forestry, had had an interest in wolves since 1966, when he and Dave
Mech radio-collared the first wolf of Dave's seminal fifty-five-year

Superior National Forest wolf study in northern Minnesota. Bob became interested in the citizen reports of occasional wolf observations in Montana in the 1970s—sightings, howling, tracks, scats, and photos—and created the University of Montana's Wolf Ecology Project (WEP). Although his PhD was in botany, Bob was a conservationist and educator who specialized in wilderness, wildlife, and environmental studies, and he mentored hundreds of future biologists. Bob would become my mentor and friend, always ready with a thoughtful nod, a wry smile, and some awful puns.

Bob's friends and students went into the field to investigate clusters of sightings to determine if there really were wolves out there. Some signs indicated that a wolf had indeed passed through here or there, with an occasional wolf found shot or run over, but the wolves had not managed to survive long enough to form packs and build a population. Wolves were scarce or nonexistent in Montana in the 1960s and 1970s. The conclusion was that there were no wolves here.

CANADIAN GRIZZLY BEAR researcher Bruce McLellan contacted Bob in 1978 with some reports of wolves in the Flathead River drainage of British Columbia, immediately north of Montana's Glacier National Park—very exciting news. Eternally optimistic, Bob hoped that the wolf would occasionally tiptoe south into Glacier. Bob hired a Minnesota wolf trapper, Joe Smith, to come out in the winter of 1978–1979 and work with Ursula Mattson, who had been out in the field investigating wolf reports for Bob. Joe succeeded in trapping and radio-collaring a light gray, female, adult wolf on April 4, 1979. Bruce and his wife, Celine, helped Joe radio-collar this wolf. They named her Kishinena (pronounced "Kish-a-NEE-na") after a beautiful creek that flowed from British Columbia into Glacier National Park, as too would this wolf. And so began the WEP field study in the North Fork of the Flathead.

The plan was that I would join Ursula (whom I would replace), Mike Fairchild (who would be my trusted coworker for many years of the project, commuting up from his home in Kalispell as the project required), a handful of seasonal WEP volunteers, and the Border Grizzly Project (BGP) crew. My job was to trap and collar more wolves for the WEP and also to trap and collar coyotes for my master's thesis on coyote territorial behavior, while alternating time in the field with time in the classroom at the University of Montana in Missoula, two hundred miles away. My learning curve was steep. My prior wolf trapping experience in northern Minnesota helped, but I still had a lot to learn.

I DROVE MY little Toyota Hilux two-wheel-drive pickup slowly to the Canadian border on a September afternoon in 1979. The sixty miles of dusty, potholed gravel road took two hours to navigate. Grass was growing in the middle of the road the last few miles. I parked at the old U.S. Customs log cabin, its hand-hewn cedar shake roof a veritable terrarium of green and white lichens. John Senger, the young, amiable U.S. border guard, had his feet propped up on the porch rail, reading *The Monkey Wrench Gang* by Edward Abbey. This was a rather relaxed government operation.

The Trail Creek border crossing was fifty miles from the three Ps of civilization—pavement, power, and phone—and there were fewer than one hundred full-time residents living within a fifty-mile radius of it. I was a long way from anywhere, in one of the most beautiful landscapes I had ever seen: a wild valley with few humans, the perfect setting for wolf recovery to begin. I had arrived at my new life to become a wolf researcher at last; I was smitten.

Across from the U.S. customs building was the log cabin occupied by the BGP crew. The BGP was started in the mid-1970s by renowned bear biologist Dr. Chuck Jonkel, who was passionate about grizzly bears, his crew, and his students. Chuck was a professor at the University of Montana and, like Bob, he mentored hundreds of people over his long career. Chuck was loved by all who had the good fortune to work with him. With his broad stature, unkempt hair, and salt-and-pepper beard, he looked a bit like the grizzlies he studied. He adopted some of their habits too, leaning against a doorjamb or a big tree and rubbing his back along it, self-scratching in delight—much like a big, old bruin marking his territory. Chuck would raid the dumpsters behind grocery stores in town and bring his spoiled loot up to the BGP crew for bear-trapping bait. The food that was outdated but not yet spoiled made for many fine meals for the BGP and WEP crews.

I WAS FORTUNATE to have Bob and Chuck on my master's thesis committee, along with Dr. Bart W. O'Gara. Bart was a retired career military man and crack marksman who was sometimes called "Bloody Bart" for his hunting prowess. I took his Wildlife Diseases and Parasitology course; he was permitted to shoot deer or elk for scientific purposes for his classes. I stood in awe as he leaned his scoped rifle across the hood of his truck, watching a group of three mule deer does bounding away across the open landscape—unhurriedly taking his time aiming, then lightly squeezing the trigger. BANG! One of the bounding does dropped to the ground and didn't move. Bart and our class walked out to the dead deer together and performed the field necropsy (equivalent to a human autopsy). Bart pointed to the bullet hole in the back of the deer's skull and lamented that he was about an inch left of center—when he had been aiming for the center. How had he even hit that deer, bounding more than one hundred yards away, and place a perfectly lethal shot to the head? He skinned and gutted the deer efficiently, and we went through every organ and tissue, looking for lungworms, tapeworms, liver flukes, and anything else that might pop up. Fascinating!

I EASED OUT of my truck at the Trail Creek border crossing, shook the road dust out of my hair, stretched, and met the BGP crew and the WEP crew who were living in the BGP cabin. I was pleased to join this small community of adventurous researchers. The BGP crew used their cabin as a base, and alternated weeks and locations on their traplines so it didn't get too crowded. The twenty-foot-by-twenty-foot cabin had a sleeping loft upstairs; downstairs was the kitchen, a living room with woodstove, and a bedroom with bunks. It was cozy indeed. The hardy twenty-somethings there were living the life I had dreamed of as they trapped, radio-collared, and tracked wild grizzlies. Fangs and fury, the real deal. Bruce and Celine, the Canadian grizzly researchers, lived a half mile north of the border across the North Fork Flathead River, commuting across by canoe.

The BGP and WEP crews kept each other informed of grizzly and wolf information. We used the canoe, or the self-pulled aerial cable car, to cross the river for a rendezvous with Bruce and Celine—or to access trucks that we would leave parked on the Canadian side of the border after-hours. While our two crews were out working our butts off, tracking bears, wolves, and coyotes, the border guards played cribbage until they were interrupted to inspect their one car per hour. We all became good friends, sharing our lives and some outrageous year-end border-closing parties on Halloween.

A quarter mile south of the customs station sat Moose City, which was not a city at all, but a turn-of-the-century homestead ranch with a haphazard collection of six timeworn cabins, three outhouses, a barn, ghosts of previous misanthropic inhabitants, and a spectacular view of Glacier National Park's majestic peaks.

An entertaining distraction at Moose City that fall was the filming of Michael Cimino's movie *Heaven's Gate*, one of the most expensive box-office bombs that Hollywood ever produced. The caravan of Hollywoodies converged on the border daily after driving up from Kalispell at sunrise, entangling anybody trying to drive south on the narrow, twisting road. Cimino had semitruck-loads of powdery, dry dirt hauled up and dumped on our Moose City hayfield runway to film a conflict inspired by the Johnson County War where horsemen galloped and shot their way through dust clouds on expensive horses.

It took a lot of gourmet food to keep the multiple members of the production crew and the many actors fed, so the cooks set up camp with huge kitchen semitrailers, plopped right between the BGP cabin and our outhouse. When the cooks finished cleaning up the dinner mess, they dumped the cooking grease and leftover soup into the ground squirrel holes behind our cabin, and then the whole movie crew drove back to town each night. Quiet returned as darkness descended—and then the night shift came out. When we needed to use the outhouse, we had to shine a flashlight on the seventy-five-yard path, from the cabin past the ground squirrel holes to the outhouse, and determine which animal's eyeshine was gleaming back at us. If the eyeshine was orange, it was a prey species, likely a deer or an elk. If the eyeshine was green, it was a predator, likely a coyote, mountain lion, or bear. And if the eyes were more than four feet above the ground and at least four inches apart, we'd know a grizzly was out there, digging up the cooks' leftovers. On those nights, we would use a coffee can in the cabin for when nature called, cursing the stupidity of the caterers. Thanks to the bear banquet left by Cimino's cooking crew, the BGP successfully trapped and collared several grizzlies within a mile of our cabin. We could sometimes hear the roaring of a newly captured and angry bear from our bunks.

The movie caterers' blunder provided an opportunity for me to get my hands on wild grizzlies and to assist some of the best bear trappers and researchers in the country. Coming directly from Minneapolis, I was thrilled to help radio-collar the biggest, fiercest animal in North America. I was able to touch a grizzly's coarse fur, smell its earthy scent, marvel at its ivory, four-inch claws, and feel the power in its enormous shoulders. Grizzlies have a muscular hump over their shoulders to help them dig and turn over rocks—and, occasionally, to soundlessly lift a whole, rotting deer out of the BGP bait truck parked in front of our cabin while we slept.

Black bears lack the grizzly's distinctive hump, a key in identifying the correct bruin species. Being a newbie in grizzly country, I had to learn how to distinguish blacks from grizzlies so I would know when I was really in trouble. Black bears in the region can be black, brown, or cinnamon, and grizzlies can be brown, blondish, or black. But only the grizzlies have those finger-length, pearl-colored claws, shoulder humps, and dished faces. Both species strongly prefer to leave you alone, but upon occasion both species have killed people, usually in defense of cubs or during a surprise, close-range encounter. Over the next few decades, I would have many encounters with grizzlies and black bears, some thrilling and some heart-stopping.

BY MID-OCTOBER, winter began to steal autumn's colors and warmth, the peaks in Glacier National Park became white with new snowfall, and puddles were glazed with ice that shattered as I drove over them in the morning. Elk, deer, and moose hunting seasons were in full swing, and hunters staked out their traditional campsites in pullouts along muddy logging roads. I mostly worked in Canada in those early years, because Kishinena, our sole radio-collared wolf, lived mainly in British Columbia, just north of Glacier National Park. The hunters and loggers were a great source of information, and I was always asking them if they had seen wolf signs, but few had.

Our WEP crew made friends with hunters Don and Phyllis Forbes, who came to the Flathead every year from Vancouver in pursuit of elk. They rescued us more than once when our truck broke down and told us where they'd seen wolf tracks. Don told us of finding a hunter-wounded, dead elk—and he saw a wolf feeding on it. He silently led Phyllis, Mike, and me down a long, steep, brushy slope, and we hoped to catch a glimpse of the wolf feeding.

The stench of the rotting elk wafted up the hill to us. Don was stealthily slinking forward when a dozen ravens exploded out of the brush twenty yards away, scaring the bejesus out of us. He froze. And then we felt his energy change as he shouldered his .30-06 rifle, turned toward us and silently mouthed the word "RUN!" We rapidly scrambled up the hill, and Don came racing up behind us, bug-eyed and looking over his shoulder. A grizzly had displaced the wolf, claimed the kill, and was tearing apart the elk. This was an important lesson: Do not have a predetermined scene fixed in your mind when there could be something very different ahead to discover. Always be aware. You are not at the top of the food chain here.

I diligently set out traps all fall to try to catch and collar a wolf. I also trapped and radio-collared several coyotes for my master's thesis project, comparing the ecology of our lone wolf with territorial coyotes. But if there were any wolves present, they remained elusive. The grizzly researchers had warned me not to leave bait in my truck while camping on the wolf trapline because it would draw in bears that were in hyperphagia—a state of intense food drive where they obsessively consumed every possible calorie before their long winter's sleep. During the fall, bears can be extremely aggressive over food sources.

I was camping in my truck with my dog Stony while running the wolf trapline. At 110 pounds, Stony was a trusty sidekick when I was out in the wilderness. Being mindful of my colleagues' warning, at night I put my bait in the front seat of the truck, and I slept in my winter sleeping bag, in the back of the open truck with Stony. I left the tailgate down so I could see my surroundings better. On my second night, I awoke to the sound of deep guttural rumbling from Stony. I sat bolt upright and, by the bright light of a nearly full moon, I saw a huge bear standing twenty-five yards from the truck, staring at me in my frost-coated sleeping bag. Stony rocketed out of the truck, and I watched a roiling mass of snarling swipes and growls disappear down the road into the shadows of the night. My zipper was frozen stuck under my chin, as I frantically hollered and whistled for my dog and tried to escape my sleeping bag. Five minutes after I assumed the bear had killed him, Stony leapt up on the tailgate and resumed guard duty, quite puffed up and proud. I slept a bit restlessly the rest of the night, but fatigue overcame worry and I finally drifted off to sleep.

It was still dark at 7 AM as I was awakened by a storm with thunder, lightning, rain, sleet, snow, and everything that the mountains could throw at me. I quickly managed to get most of my sleeping gear under cover before it got too wet, and then slept for another hour inside the cab with Stony and the bait while waiting for the storm to blow through. By 9 AM, the sun came out, the mountains were covered in a new, white, winter blanket, the bushes were graced with frosty crystals, and the truck was entombed with a thin layer of ice. The trees were like a beautiful fairy world of sparkles and ice, chaos transformed into magic.

MY ROUTINE THAT fall and winter alternated between maintaining a rapidly freezing trapline and radio-tracking Kishinena and my radio-collared coyotes. Dave Hoerner from Red Eagle Aviation, the best bush pilot ever, flew up from Kalispell in his Super Cub with skis and landed on the snow-covered hay meadow at Moose City. Mike was sharing the BGP cabin with me that season, and the two of us packed the crude runway with our snowmobiles so the plane wouldn't sink into soft snow. Flying was the highlight of the week for me. I loved seeing the North Fork from the air and catching glimpses of the animals through the dense forest canopy.

Mike and I helped Bruce and Celine collar their last bear of the season—a crotchety, eighteen-year-old female named Blanche whose three cubs of the year were running around us while we worked on their sedated mother. It was both exciting and terrifying. Blanche was a mass of powerful muscle and crucial insulating fat, having spent the summer bingeing on huckleberries, roots, and moths, and all fall feasting on hunter-wounded animals and gut piles. We also helped Bruce and Tim Thier collar a 425-pound black grizzly with immense, ivory-colored claws whom they named Pepe, after Pepe Treviño, the Mexican grizzly bear biologist that Chuck Jonkel brought to the North Fork. Pepe hoped to capture grizzlies in Mexico if there were any remaining in the wild, and he was with us to learn how from the BGP crew.

Pepe showed me a photo recently taken of a scared, wild Mexican wolf cowering in a shed in the Chihuahua region of northern Mexico. This male wolf was possibly the last wild Mexican wolf roaming free in North America, and he was looking for a mate. The wolf had been bringing deer legs to the rancher's female dog to pair-bond and start a pack. The rancher had called authorities, wanting the wolf gone. Given ranchers' hatred of wolves in general, I was impressed that the rancher had not shot the wolf. The lonely male wolf had been captured, temporarily confined in the shed where the photo had been taken, and then moved into captivity, where he likely became one of the founders of the Mexican Wolf Recovery Program. I felt a hollow melancholy, thinking of this lonely wolf hungering for a mate in wolfless country and risking his life for a ranch-dog lover. Many heartbreaking stories have been written about the last renegade wolves in the West, brutally killed by government trappers in the 1920s and 1930s, who even caught and killed a wolf's mate and used the dead wolf for bait. This was a modern version of the same tragic story, except that this wolf lived, albeit in captivity.

ALL THAT FALL I did my best to try to capture a wolf, but you cannot catch something that isn't there—or that is smarter than you are. Kishinena had been trapped before, and the traumatic event was seared into her wily canine brain; she would not repeat her mistake. My goal was to trap more wolves and to trap Kishinena before the battery in her radio collar gave out. For now, I had to content myself with getting radio collar beeps from Kishinena and to see her tracks near Proctor Lake, Nettie Creek, Sage Creek, Beryl Lake, and other places that were difficult to access.

While trying to outwit the smartest animal in the forest, I learned much about other North Fork wildlife preparing for winter. I saw a snowshoe hare whose coat was transitioning from summer brown to winter white. Chattering tree squirrels rapidly stripped cones and made huge middens of cone scales under the evergreens, stashing away seeds for hungry months ahead. White-tailed deer changed from their sleek red summer coat to a dull gray one that covered their jiggling haunches as they bounded away, fat and ready for a winter of near-starvation. A powerful goshawk skillfully dodged branches while pursuing a gray jay through a dense thicket of pines. The lucky jay escaped; the unlucky hawk went hungry.

After the canoe became useless for icy winter river crossings, I learned how to cross the North Fork in chest waders.

A river is a magical, dynamic, living being. In the summer it greets you with a cold bath on a hot summer's day. Maybe you catch your trout dinner there on a fly rod. But in winter, the cold and ice choke it down, narrowing the flow into churning channels of dark open water cutting through the ice. As a wolf tracker, you must learn how far out you can safely slide from shore on the ice shelf before it breaks out from under you. You stand there on the ice in your chest waders on a subzero day, listening to the slush ice swishing along on the river surface. You step carefully to the edge of the ice, look down into water absolutely clear to the river bottom, and find a shallow place where you can slide off the icy ledge and land softly on slippery rocks. You watch for rocks coated with anchor ice, which looks like furry slush. In fast-flowing rivers during periods of extreme cold, when the supercooled water encounters objects at freezing temperature on the stream bottom, the water will freeze to them, creating anchor ice.

You slowly navigate your way across the river with your skis lashed onto your backpack and ski poles in hand, which you use to help keep your balance as you pick along in the swirling water. As you step onto a patch of anchor ice, the glob of slush breaks free and rises to the top to join its comrades drifting down the river. Large floating pieces of solid ice bang into your thighs and threaten to knock you off your feet. You reach the other side of the river and find a place on the shelf ice where you can carefully pull yourself up without it giving way. This is a difficult moment, when a misstep can dump you back into the river upside down wearing a full backpack. You crawl across the ice shelf, reach the shore, take off your chest waders, hang them in the willow bushes, and put on your ski boots and skis. "Well, that wasn't so bad," you think to yourself, because you are young and immortal, and you ski off to find wolf tracks.

The international border station closed on Halloween, but our fieldwork continued throughout the winter. We had an official arrangement with the U.S. and Canadian customs agencies that we would keep a log of our border crossings between November 1 and June 1, when the Trail Creek customs office was closed, and we would turn in a handwritten logbook of our crossings to them the following spring. It worked well for everybody. We were allowed free access 24-7, and customs used us to report border breaches by vehicle or snowmobile; it didn't cost them anything. We were grateful for their trust so we could continue our fieldwork.

AFTER A SUMMER and fall living at close quarters with the grizzly researchers, and a winter tracking coyotes and Kishinena with volunteers, I moved out of the BGP cabin and into my own cabin in Moose City to create a more private life for myself. It was the original homestead cabin and had been built before Glacier became a national park. I became the official Moose City caretaker for the distant landowners, who would occasionally come out for a visit.

My arrival brought the permanent Moose City population to three: me and my two dogs, Stony and Max—my canine family. I had picked Stony up at the Anchorage dog pound in 1978 while working a seasonal job in Prince William Sound, Alaska, studying marine birds and mammals. If you crossed an Irish wolfhound with an Airedale, you'd come up with a close likeness of Stony. Max was a shaggy, blond, seventy-pound mongrel that I adopted from the Missoula animal shelter in 1980. He looked like the original golden-doodle, except that he only cost me $30. In addition to my family, if you counted the mice, deer, elk, grizzly bears, coyotes, and moose with whom I shared the place, the population would number in the thousands, a true wildlife mecca.

I loved my 1909 homestead cabin. Successful living was all about simplification and improvisation—and truly living in the moment. I knew where inside the cabin in the southwest corner to put the buckets when it rained heavily. I regularly stuffed the cracks between the logs with rags before the winter winds set in, and I woke up keen for each new day's adventure.

I showered in a tiny wooden stall on the porch that had an over-head, two-bucket water system. I heated water, or melted snow, on the woodstove, climbed up a stepladder, and poured the hot water into the top bucket. Then I stripped down, stepped naked onto the wooden slats in the stall, and pushed up on the edge of the upper pail with an old wooden ski, which pivoted the top pail at its han-dles so it would tip. The contents of the top bucket would then spill over into the lower bucket, which had a couple dozen holes punched through its bottom, creating a giant showerhead which poured steamy, hot water over my head and body. Glorious! I used this system until early winter when the floor slats would fill in with ice and the water wouldn't drain out anymore. Then it was bath-in-a-basin, indoors, for the rest of the winter.

Refrigeration was provided by nature and my snow-filled cool-ers until I upgraded to an ancient propane refrigerator out on the porch. To make the refrigerator cold, I lit a flame underneath it, let-ting the ammonia, water, and hydrogen gas in the tubes percolate, evaporate, condense, and do their cooling thing. Pure genius. I cooked on an ancient, cast-iron stove heated with wood, and sometimes I prepared a meal on a two-burner Coleman stove when I was in a hurry. I eventually upgraded to a small, four-burner propane stove. What a luxury to heat up spaghetti instantly, without having to build a fire in the cookstove and slowly warm up the whole cabin in order to cook dinner.

Light was provided by three propane lights that I installed above my dining table, which doubled as my desk. When reading or writing up my field notes, I supplemented the dim propane lights with a Coleman lantern. Many evenings I sat at the table with a pencil and field notebook, filling in data sheets to the hiss of the Coleman lantern, as my dogs snored on their pile of blankets.

I also found time to do oil paintings of wildlife and landscapes, especially when the weather was too inclement to do fieldwork. I sat cross-legged, mixing greens and blues and browns (typical North Fork colors) on an enamel tray, and painted the background for a fox, moose, or harlequin duck on the stretched canvas with patient brushstrokes. Sometimes I was so caught up in creating a painting that hours passed without me even realizing it—that is, until I tried to stand up and my legs were tingly and cramped. I sold my oil paintings and pen-and-ink note cards to help support my wolf habit.

I fell head over heels in love with my new digs and my independent lifestyle, and the study of the recolonizing wolf population in the North Fork became my world for the next eighteen years. I had the Moose City owners' generous permission to house the WEP crew in some of the other cabins in Moose City when there was funding and work. After the border closed and before the seasonal volunteers arrived—and before 1985, when the WEP started up again after a three-year hiatus in funding—Stony, Max, and I usually had the place to ourselves.

ONE DAY I was cross-country skiing up Whale Creek Road when I found a boot-stomped trail in the snow, leading off from some tire tracks on the side of the road into the forest. I was curious what this person was doing, so I took off my skis and punched along in his boot prints. I walked seventy-five yards into the woods and heard a strange sound, part growl and part scream. And then I saw a wretched pine marten hanging by her left front foot, which was caught in a trap set along a horizontal tree branch. A wild pine marten is one of the most agile and cute little furbearers you'll see in woodlands and conifer forests. They look like a cross between a mink and a small, skinny cat, with perky little ears, an orange throat patch, and a luxurious sable pelt. Don't let that cuteness fool you, however—they are miniature wolverines.

This ill-fated marten had smelled a trapper's bait and run up the tree trunk, along the tree limb, and over the top of the trap, which had snapped shut on her leg. The panicked marten launched off the branch to free herself of the thing clamping her leg, only to hang suspended from the branch by the steel shackle. She was dangling from the branch as I approached. Her leg was mangled, swollen, and frozen, and her face and shoulders were covered with frost. She screamed out of anger and anguish and twirled in mid-air. I wished I could have released her, but her fate was sealed. Her leg was destroyed, and she had been hanging from there for God knows how many days. I picked up a large chunk of wood, gritted my teeth, and bashed in her beautiful little head with three hard blows. Anger and sorrow poured out with each whack. I removed the trap from the dead marten's leg, ripped the trap off the tree, and threw it as far as I could. I carried the marten out to the road and skied a quarter mile farther down before I tossed it as far as I could throw into the deep snow, so the trapper wouldn't find it.

I skied and drove along Whale Creek Road, and then back down along the main North Fork Road looking for the trapper's trails. I found and snapped shut seven more traps, leaving them hanging in place, disarmed so they couldn't catch any animals—to let the trapper know that I was onto him. I was trying to stay within the law, although I was pushing it—I had to live in this community for many years to come and wanted to avoid rubbing people the wrong way. But the screaming marten had pushed me over the edge.

The next morning, I was inside my cabin at Moose City and heard somebody honking frantically at the border station. I put on my skis and set off with my dogs to see what the trouble was. That trouble was waiting for me next to his pickup, which was parked by the BGP cabin across from the U.S. Customs station. It was the angry trapper, with a loaded rifle in his gun rack, and buckets full of bait, bones, and traps in the back of his pickup.

"Who lives in this cabin?" he angrily demanded.

"Nobody," I replied.

"You're lying. I see gear and crap on the porch. Somebody lives here who's messed up my trapline and I want to know who it is."

I was getting worried, but then it dawned on me that this tough, old trapper couldn't fathom that a young, blond woman might have the skill or knowledge to find and dismantle his trapline.

"Tell you what, if you leave me your name and phone number, I'll call you if I find out who was here and bothered your traps," I told him.

He wrote out his name and Kalispell phone number with a pencil on a torn envelope and handed it to me. He was cooling down now; he went around to the back of his truck and rummaged through his bait pile.

"Here's some bones for your dogs," he said, as he grabbed some animal legs for my always-hungry dogs. He patted each dog on the head as he gave them each a bone. Then he drove away. I breathed a sigh of relief but smiled and thought that it takes a trapper to catch a trapper. And he had missed this perpetrator.

I LEARNED MANY lessons that year: Be mindful of your surroundings and they will teach you lifelong lessons. Be careful; there are things out there that can kill you. Cross the icy river in waders in January with caution; you only fall once. Don't shoot pack rats inside your cabin or the ricocheting bullet might get you, too. Be a good teammate and your pack will help you. Nature always wins.

Our research that year would confirm the return of wolves to Montana through natural dispersal, fifteen years prior to reintroductions into Yellowstone National Park and central Idaho. What unfolded over the coming decades is an untold story of how wolves returned to the West, unaided by humans, but intensely monitored by the enthusiastic and gritty members of the Wolf Ecology Project. The big credit goes to a handful of extremely lucky colonizing wolves who would become the founding members of dozens of packs throughout western Montana, Idaho, southeastern British Columbia, and southwestern Alberta in approximately half my lifetime: Kishinena, our first collared wolf, began a wave of wolf recovery that would result in more than a thousand wolves in Montana and nearly three thousand wolves throughout the western states in just forty years. Most importantly—for me, at least—was that I was home at last.

4

....

KISHINENA

THE NEW YEAR rolled quietly into my cabin, with the snow sifting in under the one-inch gap of my frosty Moose City door. Overnight the temperature plummeted in the clear, starlit sky. It was so cold that sugar-rich sap froze and trees exploded like rifles, waking me up inside my cabin several times. I have a superstition that how I spend New Year's Day sets the tone for the rest of the year, so I spent the first day of the year doing something I love— skiing along wolf tracks.

I made a thermos of hot tea, slathered together two peanut butter and jelly sandwiches and put them into ziplock bags, packed up my gear, and headed outside. After checking in the toolbox for an extra spark plug, tools, and rubber belt, I pulled the starter cord repeatedly on the old snowmobile until it sputtered. Stinky, blue clouds of oily exhaust poured out of the engine, and I feathered the gas thumb throttle until it settled into a smoother roar. I lashed my cross-country skis, poles, tracking gear, backpack, lunch, extra mittens, warm clothes, thermos, and insulated water bottle to the back of the seat, and I roared north past the closed Canadian border station in search of Kishinena.

The snowmobile wallowed through the three feet of snow that filled the abandoned logging roads. Periodically I stopped, killing the ring-ting-ting of the obnoxious engine, and reveled in the absolute silence for a long while. I unpacked the telemetry antenna and receiver to listen for the beeps of my radio-collared coyotes—and

Kishinena. I thought to myself, "Pause, swing the antenna slowly in an arc while listening carefully, fiddle with the receiver dial, turn 180 degrees, pause, slowly swing the antenna, and listen." A chickadee called from a nearby spruce, but otherwise all I heard was the blessed sound of silence. I carefully repacked the fragile antenna and receiver, then repeated this entire process at two-mile intervals until I began to pick up the radio collar signals. What a thrill to be seeking the signal of the only wolf regularly visiting the Lower 48 west of the Mississippi River, and nearly the only wolf in this remote corner of Canada. Studying this first intrepid coloniz-ing Flathead wolf was like arriving on Isle Royale as the first two wolves crossed over the frozen surface of Lake Superior in 1948. I was privileged to be here as wolf recovery unfolded.

I DIDN'T HEAR the first ping until I stopped at the Upper Sage Creek Road, where Kishinena's signal began to come in faintly. I snow-mobiled north up the road. Three miles later, I saw Kishinena's fresh tracks in the snow on the north end of a bush airstrip built by an oil and gas exploration company. Her signal was much louder here. I shut off the snowmobile, unlaced my insulated Pac boots, pulled on my three-pin ski boots, unstrapped my skis and poles, and began to ski forward, following her tracks in the direction she had traveled.

She had trotted through the woods to a bluff above Sage Creek, then turned back toward the airstrip. After two miles of meander-ing, her tracks passed along a chain of marshes with three frozen beaver dams. She paused occasionally, as if detecting a faint scent of something tantalizing. She encountered tracks of a single small moose on the last beaver pond, and she followed them, her pace picking up. I pictured her ears tilted attentively forward, her nose lifted into the breeze, her muzzle keenly shifting back and forth, and her head bobbing slightly, zeroing in on the prize ahead. A hun-dred yards farther on, her trotting tracks merged with the ambling tracks of the moose browsing on willow twigs, both animals travel-ing westward through the lodgepole pines. Then the stealthy wolf broke into a run. Eighty yards ahead, the pattern of the moose tracks changed from a walk to a trot—and then a gallop—as the animal detected its pursuer.

Wolf and moose tracks mingled in a torn-up trail of predator and prey as they both ran for their lives, one out of fear and one out of hunger. Large clumps of moose hair appeared along the trail where the wolf had tried to grab onto the fleeing moose. After two hundred yards of skirmishing, broken branches littered their trail and a few blood spots dotted their entwined tracks. Then the wolf tracks disappeared altogether from the chop and churn of run-ning tracks. Where did she go? What was happening? Thirty feet later, the wolf tracks reappeared in bursts among the moose tracks. The moose was slowing down. And seventy-five yards later, I came upon the freshly killed moose calf with wolf tracks ringing it.

I skied up to the female moose and saw a ripped-open slash just behind the ribs that was literally steaming where the wolf had begun to feed. The kill was so fresh that the moose's eyes hadn't yet begun to glaze over. The warm body was barely fed on; the steam-ing guts were slightly protruding through a hole by the ribs, and a few chewed rib bones indicated the wolf had begun opening up the moose here to eat. I examined the warm carcass, tugging at the dead weight until I was able to roll it over to examine all sides of it for injury or disease. I was surprised to find nothing wrong with the moose other than the eviscerated hole and some pink, nickel-sized bare spots on its neck. I was thrilled to have such a fresh kill to investigate.

After more thoroughly examining the chase scenario and the dead moose, my excitement ebbed away as it dawned on me that I had disrupted Kishinena's hard-earned banquet; I felt terrible about scaring her off. But I figured it was only a temporary inconvenience to her. Heck, I had just found Kishinena's first-known kill, I was already there, and I had data to collect. I pulled out my folding hunting knife, ziplock bags, and tape measure and knelt in the snow over the moose. I carefully cut into the skin near the pink marks and peeled back the neck skin. Beneath these marks was massive hemorrhaging, and there were purplish blood clots along the layer of muscle near the windpipe and arteries.

Then I put the story together: Kishinena's tracks disappeared along the moose's trail because the wolf had leapt up, grabbed the moose by the throat, and hung on—trying not to be trampled by flailing hooves as the moose thundered forward, head tilted up high, lifting the clinging wolf above the ground. Eventually, the moose calf succumbed to the crushing of its windpipe as its oxygen supply was cut off. Kishinena's teeth were apparently not sharp enough to puncture the moose's thick skin as the wolf hung on with her canines, but the sliding of her fangs over the moose's hide scraped hair off and created the pink, nickel-sized bare spots. Her bite pressure, roughly one thousand pounds per square inch, inflicted severe damage under the skin and effectively strangled the moose.

The fact that Kishinena had taken the risk of tackling a prey animal four times bigger than she was, by herself, made me marvel at her strength and will to live. I took a marrow sample from the femur, or thighbone, to evaluate the moose's overall body fat and condition. It was gelatinous and red, indicating that she was starving and in poor shape. There were no tracks of any adult moose defending this lone calf from a predator, making the orphan a perfect target for Kishinena—and a bonus for a hungry graduate student. Seeing the mountain of fresh moose meat, I sliced off five pounds of warm backstrap steak, put it in the bottom of my backpack next to my bagged samples, and skied out to the snowmobile. I enjoyed fresh moose steak dinners for the next three nights and thanked Kishinena for our shared bounty.

FOUR DAYS AFTER Kishinena killed the moose and one day after I finished off the moose steaks, I skied back to the kill site. I was looking forward to seeing how much of the moose calf Kishinena had consumed. On my way to the site, I saw four-day-old tracks where Kishinena had encountered my ski tracks. Here she had stopped abruptly, paused, sniffed, and then turned around and headed back in the direction she had come from. I thought it peculiar that she hadn't crossed my ski tracks. I assumed that she must have gone around and returned to her kill from a different direction. I skied on toward the cackling magpies and gronking ravens to where I found the almost completely consumed moose carcass. I was feeling really good about Kishinena enjoying her feast.

Looking more closely, however, I realized that the only fresh tracks around the moose carcass were those of coyotes and scavenging birds. Kishinena's tracks from the original kill time were faintly visible under that trampling of the freeloaders. I cautiously skied around the remains, examining my old ski tracks and Kishinena's tracks, searching desperately for fresher wolf prints. The sad realization hit me that after I jumped Kishinena from her hard-won kill, she was too spooked to return to it. I felt deep regret that I had ruined her banquet and cost her thousands of precious food calories. My thrill of the kill had turned into a remorseful lesson. From then on, the WEP crew would only ski on wolf tracks in the direction they came from, backtracking instead of forward-tracking them. No more disturbances like this one.

We later learned that Kishinena's avoidance of humans was remarkable. Indeed, future generations of wolves would use our ski tracks and snowmobile trails as a highway to travel to places more quickly. Perhaps Kishinena's aversion to humans was a survival skill that helped her to become the first successful colonizer to travel hundreds of miles down from Canada. She learned how to run that gauntlet by being smart and avoiding being seen by humans.

That winter, Mike Fairchild and I searched for Kishinena's radio collar signals daily, and once a week Mike and I took turns flying with Dave Hoerner in the Super Cub in search of the wolf. Unfortunately, her radio collar ceased transmitting only a year and a half after she was collared, making it impossible to track her via her signals. However, we skied and snowmobiled hundreds of miles searching for her. We learned that Kishinena roamed throughout the North Fork alone, investigating beaver lodges, deer wintering areas, coyote urinations, river bottoms, and mountain ridgelines. We continued to see tracks of a female wolf, as revealed by the urination pattern in the snow, in Kishinena's territory, on her usual travel routes, throughout 1981. It had to be her.

I WAS EXCITED in early 1982 when Glacier National Park rangers Jerry DeSanto and Steve Frye came to my cabin in Moose City to tell me of a pair of wolves traveling together nearby in the park. "C'mon Diane, let's go check them out," they said. We waded the North Fork Flathead River, put on our skis, and headed out toward the Kishinena Patrol Cabin five miles to the southeast. They showed me Kishinena's tracks; she was traveling with a male wolf who had a unique signature—he was missing one toe on his right front foot. The wolf courtship was displayed in their bed sites—where they curled up next to each other—and in their paired urinations in the snow, with his raised-leg urinations and her squat urinations marking the same spot. They had pair-bonded! For the remainder of the winter, we skied and snowmobiled scores of miles hoping to cross the new wolf pair's tracks—not an easy task with no radio collars and a lot of wild country to cover. Knowing Kishinena's travel routes helped us locate the budding pack, and we were thrilled to find that the pair stayed together through the winter breeding season and early spring. As the snow disappeared, we lost their trail and hoped we would get sightings—and find a litter of pups.

One hot day in June 1982, the three-toed male was accidentally captured in a snare set by the grizzly bear researchers. Bruce McLellan and Dan Carney came down to Moose City to get me to help with processing the wolf. When we drove up to the capture site, the handsome black wolf was frantic and panting heavily. We sedated him as he hopped around in the sun, secured by his front foot in the cable snare. We checked him over carefully—everything looked okay. We radio-collared the wolf and released him on-site, but the following day we heard a mortality signal coming from his collar. When a wolf hasn't moved for four hours, the pulse rate of the radio collar doubles as it switches to mortality mode—and you can presume the wolf is dead.

We found his body not far from where he had been captured. His cause of death was unknown, but I suspect it was the heat and the stress of capture. My heart sank. The presumed pups would now be fatherless. At about seven weeks old, they would be in need of a lot of food and protection. I was worried that Kishinena would be unable to kill enough prey to feed them, and concerned that there would be no adult wolf to safeguard them when she went off hunting alone.

The ultimate proof came when seven pups, four gray and three black, were seen in July on a logging road six miles east of where Kishinena had been radio-collared. This first documented reproduction of wolves in decades happened eight miles north of Glacier National Park. But would the wolves make it through the fall hunting season in Canada, where they might be illegally shot by hunters?

KISHINENA WAS PLUCKY and lucky, and all seven pups survived through that fall and winter, as evidenced by the eight sets of tracks seen traveling together. We finally had a wolf pack to follow on our skis, instead of a solitary wolf. We don't know the ultimate fate of Kishinena; she simply disappeared. But her progeny lived to reproduce and repopulate the North Fork and nearby wildlands. The shy and resourceful Kishinena was the springboard that launched wolf recovery in Montana and the West. She succeeded with good instincts and without human fanfare simply by being smart and resilient.

Kishinena and I had traveled different but parallel paths to arrive at the same place at the same time. One North Forker told people that I brought the wolf with me from Minnesota and dumped her out. Never mind the fact that Kishinena had been radio-collared in the North Fork half a year before my arrival. Others said that these first wolves were remnants of a dog-sledding team that got away and went feral. I don't know of any amber-eyed sled dogs that stand deer-height at the shoulder and can pull down a moose by themselves. Both Kishinena and I were solitary females, dispersed from our natal packs, who traveled a long way from familiar home places and did our best to survive in the wilderness of the Flathead—and both of us learned to thrive in our chosen new territory.

5

...

LOGGER JUSTICE

WOLVES AND MONEY were scarce in 1981. I was finding my path through life at Moose City and the North Fork, and doing my wolf and coyote fieldwork on a paltry graduate student stipend. The pine bark beetle had come through and killed thousands of acres of lodgepole pines in 1979, carpeting entire mountainsides in dying, red-needled trees. It was a heyday for the local logging operations, which came in to salvage this standing timber before the trees began to topple over into Mother Nature's biggest game of pick-up sticks.

Border relations between Canada and the U.S. were friendly in those days, allowing Canadian logging companies to harvest trees north of the border in British Columbia, haul the logs through the Trail Creek border crossing a quarter mile north of Moose City, and then unload them onto a vast, open log deck behind the U.S. Customs station. From there, a Montana company would load the logs onto their U.S. trucks and haul them sixty miles south to mills in Columbia Falls. It seemed a shame to me that Canadian trucks couldn't drive the logs straight to the Montana mill, but international relations only go so far.

Encountering these enormous, loaded logging trucks flying down the gravel North Fork Road was a common and terrifying occurrence as I drove around tracking canids in my little pickup. One of the log truck drivers told me that it takes a quarter mile for a loaded truck to stop while hauling a long bed of heavy logs. I rode along in the cab of a log truck one day to see what it was like. I remember seeing the long glass window at the back of the cab, one foot behind my head, with the butt ends of logs swinging as we rounded the corners. I drove my truck cautiously because being unlucky meant being dead.

Over time, I got to know several of the truck drivers as we passed on the road or stopped by the customs office for coffee or paperwork. I discovered that they didn't want to run over me—it would ruin everyone's day, besides making them late on delivering their loads. They would radio ahead to each other when they spotted me so their buddies could avoid squashing me. When they found out that this twenty-five-year-old, lanky, blond girl worked alone trapping and tracking wolves and coyotes, their respect for my tenacity seemed to go up a notch or two. They didn't love wolves or biologists, but at least I was a novelty in their day of hard work.

Some of the drivers were friendly and waved at me, and some did not. Most of them came from a bloodline of loggers and mill workers. Environmental regulations, road closures, and protections for endangered species were viewed as threats to their livelihoods and families.

And it wasn't just the loggers who didn't like endangered species mucking up their sources of income. Some staff in the federal and state agencies had a bellyful of grizzly bear work restrictions, since grizzlies were protected in 1975 as a threatened species under the Endangered Species Act. Despite the grizzly bear listing, Montana permitted grizzly hunting until 1991 under an exemption to the federal protections that allowed fourteen bears to be killed each fall. But the Forest Service had to address grizzly bear presence with federally required environmental reviews and other paperwork that reduced the timber cuts—and thus reduced income—in the North Fork and other national forests in northwestern Montana.

I took WEP's quarterly report into the local Forest Service office in Columbia Falls to give to the district ranger. I introduced myself respectfully and handed over our WEP report documenting Kishinena's movements: a couple of location dots in his district and the rest in Canada. He looked me in the eye sternly and said, "Thank you, but there are no wolves here." I replied, "Oh, but there is a wolf—who spends most of her time in Canada, but sometimes walks down here into your district." He set the report on the giant pile of papers on his desk. I thanked him for his time and left his office.

The district ranger may not have liked having to accommodate another endangered species in his district, but I was happy to learn that some of the log truck drivers took an interest in the wolves. Bob was at least twice the age of the other drivers, and he and his son were both hauling Canadian logs south from the border. Bob adopted a grandfatherly attitude toward me, and we sometimes sat chatting through our respective truck windows—me craning my neck up to look at him in his cab, he smiling down on me. One morning, he flagged me down. I got out of my pickup and climbed up the step to his door so we could talk at eye level, and he pulled out some photographs he had taken the previous week. I marveled at the clear images of a large gray wolf crossing the road about two miles south of the border. Great information for my research. So began our conversations about wolves returning to the North Fork. I was pleased to hear Bob describe the excitement he had felt at seeing his first wolf and being lucky enough to have had his camera handy. He liked the thought of these wolves living here. What a refreshing viewpoint from a log truck driver. Whenever we had the opportunity to chat, I would give him the latest update on the wolves. It made both of our days better.

ONE COLD NIGHT later that winter, I had gone to bed upstairs in my rustic and isolated Moose City cabin when Stony and Max began growling, and then went roaring down the stairs in full protection mode. I peeked out my bedroom window and saw the moonlit figures of two men staggering behind the corner of my garage with booze bottles in hand. I slowly descended the stairs, then turned out my highly agitated dogs to drive off the intruders, sending them outside with a snarling "get 'em!" But the wind was wrong and my dogs went tearing out into the meadow, barking and searching several hundred yards beyond the two men. Damn. Now what?

My .30-06 rifle was sitting next to the door with unchambered shells in the magazine. I had killed deer and elk with it and was familiar with its power. I picked up the rifle, quietly opened the door, pointed the end of the barrel at the men now thirty yards away, and loaded a round into the chamber. The bolt slapped closed with a resounding clunk.

"Step out where I can see you," I demanded.

The intruders stepped nervously around the corner of my garage, becoming easy targets in the bright snow.

"Hey Diane, we were only funnin'. We brought you something to drink, so let us come in."

I had no idea who they were, and the fact that they knew my name was unsettling. I was absolutely calm, scary calm, looking at them down the barrel of my rifle, the butt wedged into my shoulder. Like a mountain lion crouched up on a big rock, with steely-eyed focus and not a twitch, watching an unsuspecting deer approach. Whether or not there was a kill depended on the actions of the prey.

"You leave right now and don't come back," I heard my calm, lowered voice say. I clicked off the safety so it was ready to fire.

"Okay, okay. We're leaving now. Put down your gun!"

I stood stock-still, keeping the sights on the center of the chest of the bigger guy. They stumbled their way back toward the border, where they had left their pickup, and I watched them down my barrel until they were out of sight. After my dogs returned from their fruitless search in the meadow, they finally caught the men's scent by the garage and got all worked up again, but I ordered them to come back into the cabin.

Only then did I succumb to the adrenaline rush, and I began to shake and then shake harder. Tears flowed down my cheeks at what might have happened, how a complete coldness that I had never previously experienced had gripped my brain and behavior. That focused iciness was as frightening to me as the appearance of the two men. I hugged my dogs and tears dripped into their fur. What if I had shot the men? Life in prison? Or what if they had raped me? I drank chamomile tea for a couple of hours until my fear faded.

Did I overreact? What just happened and why? Did I just create a whole new sport for passing truck drivers who knew that I lived alone? That I could be snuck up on in my cabin—and might end up a victim—was something I had never before considered.

The next morning, I was at the border in my pickup as Bob was arriving for a load of logs. He stuck his arm out the window and signaled me to come over, that he wanted to talk. I walked up to his logging truck and said good morning.

"I heard you had some visitors last night." I looked up at Bob's silver hair and kind blue eyes, but I didn't say anything.

"Don't worry," he said. "They won't bother you again." He gave me a knowing smile, waved softly, and put the log truck into low gear; the big truck rumbled as he pulled away.

I can only imagine what stories went over the log truckers' CB radios that day, but I think Bob was my guardian angel. I never saw those two men again. Someone told me exactly which two truckers had made the midnight visit to my cabin. A few months later, I read in the local newspaper that one of them had been charged with the assault and rape of a Kalispell woman. I recalled my icy calm reaction that haunting winter night and was thankful that I hadn't become a victim too.

LATER THAT SAME winter, the only knot bumper walked off the job at the big log deck behind the U.S. Customs station. His job was cutting loads to the right length and trimming off branches and knots so the logs fit on the American logging trucks. Without him, the log-hauling process came to a halt. The boss, Bill, had a lot riding on completing his contract, and he was not happy. They were nearly done for the season, and he had zero prospect of picking up a new guy for only a couple weeks of work. Bill tentatively asked me if I was interested in a little hard work bumping knots for a few weeks for good money.

"You know how to run a chain saw?" came the logical question.

"Of course!" I said. Not quite a lie. I had learned how to use a chain saw that fall while cutting my firewood.

"Great. Show up at the log landing tomorrow morning at 8 AM. I'll provide the chain saw."

"Thank you. I'll be there."

I might have stretched the truth too far, but I was committed now, and I badly needed the money. I showed up a few minutes early, and Bill handed me a huge, battered, blue and silver chain saw that was older than I was. And really, really heavy, because it was made of all metal parts. My orange Husqvarna was mostly plastic, a lot smaller, and about one-third the weight of this behemoth. Not only did I struggle to pick it up, but the cold morning air made the compression so tight that I couldn't pull the cord out to start the monster. Bill was a good man and needed this work done; he could see that I was willing but needed a little help. So he pull-started the saw for me and told me to try and keep it warm throughout the day, which would help soften the compression. Yessir!

Then he introduced me to Jackie, the skidder operator, who had no desire to deal with some blond girl on his log deck. After the introduction, Jackie spat out a little stream of tobacco chew, brown juice dribbling down his chin. I was intimidated but kept on walking toward the mountain of logs. Bill left us to our work. I sure as hell couldn't ask Jackie to start my saw for me, so all day long, whenever the saw began to cool off between loads, I had to repeatedly pull-start that saw and keep the engine warm.

Partway through the morning, I had a question for Jackie, who had left his skidder running and was nowhere to be seen. I came around the front end of his rig, and there he was—with his tool out of his trousers—taking a leak. We saw each other at the same moment, and he actually growled at me like a bear as he zipped up his fly. I was mortified. But by the end of the day, I had learned how to bump knots, keep the chain saw warm enough to pull-start it, fuel it as needed, and stay out of Jackie's way. After all the logs were loaded for the day, I wearily walked the quarter mile back to my cabin. I could barely uncurl my fingers, which had been gripping the heavily vibrating saw all day long. I was exhausted, but I had survived Day One. Could I do this for another two weeks?

EACH DAY BECAME easier as I got used to the routine, figured out shortcuts, and gained strength. The log truckers would drive their empty rigs into the log deck, and while they waited for Jackie to load up the logs I had trimmed, they would tease me a bit. But the comments grew more respectful when I was still there doing that miserable knot bumping job day after day, not complaining. I think everybody was surprised, including me, when we finished the job in seven days, instead of the fourteen Bill had expected it to take. And just in time it turned out—as winter was closing the backcountry roads and logging operations in Canada.

After we loaded the last log and the final truck drove off, Bill walked up to me, shook my hand, and thanked me with a hearty smile. He paid me generously, the best money I had ever made for a week's work. A couple of hours later, when I was home warming up inside my cabin, I heard a noisy diesel engine rumbling into my driveway. I stepped outside and saw Jackie in his skidder delivering a massive load of culled logs and piling them beside my woodshed. It was enough firewood to last me all winter. I just had to cut it up to woodstove-sized lengths, with my sweet, little chain saw. I suspect it was Bill who ordered Jackie to bring me the gift of firewood, but I will never know which one of them made that benevolent decision. Either way, I was delighted and surprised, and my confidence went up a notch.

6

...

SAGE

YES, THERE IT was again, a wolf howl from just outside my
cabin. I slowly opened the door and stepped outside. On the
opposite riverbank stood a huge gray wolf peering at me
from the snow-laden willows. I stared back, mesmerized, and we
looked at each other for what seemed like a very long time. Then,
the wolf slipped back into the willows and was gone.

That first winter after Kishinena whelped seven pups, the WEP
field crew saw the tracks of eight wolves along the frozen creeks
and snowy game trails. We named them the Magic Pack, because
it was magical that they had survived. The wolves remained elusive
and avoided people until one of Kishinena's gray pups grew up into
the handsome and curious yearling who paid this "howdy howling"
visit to my cabin that memorable day in November 1983.

For the next month, this wolf hung around my cabin, trying
to get close to Stony and Max. I could tell from the wolf's urinations in the snow that it was a male, and I was afraid that he would
make a quick meal of my dogs. But one morning when they were
outside, I heard Stony barking and looked out to see that the wolf
and the two dogs were only a few feet apart. The dogs were guard-
edly holding their ground while the big wolf stood playfully erect,
slowly wagging his long, sweeping tail—then dropping his chest
into a play bow in the snow. I realized by his behavior that this naive
animal was very likely an adolescent wolf exploring life on his own.
The wolf trotted across the meadow, looking invitingly over his

shoulder, and when the dogs began to follow him, I whistled them back and brought them inside.

BY THAT TIME, Kishinena had disappeared, and we had no other radio-collared wolves in the North Fork—a major detriment to our fledgling wolf research project. In August 1984, I was working up in a fire lookout tower for the summer since there was no funding for the WEP. Bob Ream was happy to have a Wisconsin wolf biologist friend, Dick Thiel, set out traps to catch and collar a wolf. Dick worked his trapline magic and caught a 110-pound, two-year-old male gray wolf six miles north of Moose City. Dick had gone back to Wisconsin that same day, so Mike Fairchild volunteered to check the traps. Lucky Mike! He radio-collared this handsome wolf and named him Sage (8401) for the creek drainage he was captured in. This was almost certainly the same wolf that had visited me at Moose City the previous year: right sex, right color, right age, in a land with only a handful of wolves. Sage clearly had itchy feet and was searching for his own territory. Moose City appeared to be the epicenter of his explorations.

The success of the Magic Pack and radio-collaring Sage turned out to be the beginning of a new era for wolves in the North Fork and led to the revitalization of the WEP in 1985. Mike and I expanded the geographic scope of our research project to follow Sage's long extraterritorial forays and the new packs that were springing up in our study area.

I captured and radio-collared ten wolves between 1985 and 1987, and our research was on a roll. Moose City became headquarters for the WEP, the only wolf research project in the Lower 48 outside of the Midwest—just me, Mike, and a handful of volunteers. These dedicated wolf trackers were critical to our research efforts, and we became a real team, sort of a pack of our own. Our intrepid pilot, Dave Hoerner, replaced the Super Cub with a more powerful, four-passenger Cessna 182 that had longer-distance flying capacity, an asset to our expanding research project.

On October 20, 1985, I found the skinned and decapitated carcass of a wolf two miles north of where Sage had been captured. I felt sick when I thought it was Sage, but a thorough search with the telemetry gear told me that Sage was alive and well in other parts of the North Fork. Five days later, I had a date with Sage; I trapped him on October 25 and fitted him with a new radio collar. When we recaptured a wolf and the collar had a year or more on it, we always replaced the radio collar to give it an extra year of life. Sage was the first wolf I had trapped since joining the WEP, and with his large size and massive head, he was magnificent. I had finally done the nearly impossible—captured a wolf in a land where they were still rare—and I felt validated that I was truly on my way to becoming a competent wolf biologist in this wild landscape, with endless possibilities ahead for research.

AS WINTER PROGRESSED, Sage wandered widely over a huge area in northwestern Montana, southwestern Alberta, and southeastern British Columbia. During this time of intensive exploratory forays, we located him dozens of miles between successive locations. Sage taught us that neither raging rivers nor the Continental Divide could impede a wolf, even in the dead of winter, in search of space, food, and a mate. On our telemetry tracking flights in November 1986, we saw Sage traveling with a smaller black wolf, which we hoped was a female.

This newly formed pair dispersed to the Wigwam River area and set up a new territory. In the summer of 1987, while radio-tracking from the Cessna, I was delighted to observe Sage and his black mate relaxing at their den with five pups. The Wigwam Pack fared well in this remote river drainage of British Columbia. The wolves had plenty of elk, deer, moose, and bighorn sheep nearby to fill up their bellies. Competing predators in the valley included mountain lions, grizzlies, and one determined hunting guide and outfitter—Henry. In early October, one of Henry's hunting clients legally shot one of the Wigwam pups. Wolves could not be shot legally in the

Flathead drainage, but they could be shot legally in the Wigwam drainage. We received a report from the British Columbia Wildlife Branch that a large gray wolf in the pack, undoubtedly Sage, had also been wounded by the hunter, but escaped. During our next telemetry tracking flight, I observed Sage favoring his right front leg. A couple of months later, he seemed recovered from whatever his injury had been, and I breathed a sigh of relief.

The morning of the last day of the year in 1987, it was fifteen below zero (Fahrenheit), clear, and calm—perfect conditions to conduct a telemetry flight and track the four packs in which we had radio-collared wolves. Our pilot Dave and I located the three packs currently living in the North Fork, and we headed northwest to the Wigwam. We observed Sage lying down on a snow-covered gravel bar along the Wigwam River. But something didn't look right. The snow was trampled and the willows were torn up. I asked Dave to make a low pass over Sage, and only then did Sage stand up and struggle to move away. His right rear leg was held fast in a leghold trap attached to a log by a three-foot chain. To my horror, as we helplessly circled Sage, he headed for the partially frozen Wigwam River. He plunged into the water, swam across the river with the trap, hauled up on the other shore, and headed into the willows— the trap and anchor log leaving a snowy furrow in his wake.

Sage had been trapped a quarter mile north of Henry's back-country camp. Henry had no love for the wolves, which he viewed as competition for his livelihood of big-game hunting. Dave swung the plane over the cabins, lodge, corrals, and outbuildings, looking for evidence of life and a place to land the Cessna on skis. The snowed-in tracks around the camp indicated that Henry hadn't been there for several days, and we had no idea when he would return. Dave told me to hang on, as he tried to land between the buildings and not slide into the river. He dropped the plane down through the camp, the wingtips missing the buildings by inches, but despite Dave's incredible bush pilot skills, he ran out of flat terrain and had to pull back up hard—just before the looming trees along the riverbank.

"Head for Polebridge," I directed over the crackling headset.

It seemed like forever until the old Polebridge Mercantile and a handful of cabins came into our view. There wasn't a runway at Polebridge, but it did have the only telephone for twenty-five miles. Dave skillfully landed in a snowy field beside the Mercantile. I ran to the telephone outside the store to call Mike—and then realized I didn't have any change in my pockets. I burst into the store and asked Betty, who was tending the counter, if I could borrow a dime for the pay phone. She opened the till, handed me a dime, and didn't ask a single question, as if planes landed in her yard regularly to use the phone. I called Mike, who was in Kalispell getting our truck and snowmobiles repaired (of course they were broken down), and he said he would have them ready to go by the time I got there—thank you, Mike! Dave and I flew back to Moose City so I could gather up the gear, clothing, and capture equipment that we would need to free Sage. Then we flew to Kalispell, where Mike picked me up, and we drove north in our truck with the gear, mobile phone, and snowmobiles.

I HADN'T THOUGHT about how difficult it would be to contact customs officials, veterinarians, and the appropriate Canadian wildlife officials on the afternoon of New Year's Eve. It was necessary to get permission from the Canadian officials to release Sage, because he had been trapped in British Columbia by a licensed outfitter and trapper. We had worked cooperatively with the Canadians for several years, and we didn't want to create an international incident. If we botched this politically, it could jeopardize the future of our wolf research efforts in Canada. To further complicate things, although wolves could be shot by trophy hunters, there was no legal trapping season on wolves, so Henry broke the law when he trapped Sage behind his cabin. Henry loved biologists as much as he loved wolves.

We called a veterinarian and received detailed instructions on how best to thaw a frozen foot and deal with the thermoregulatory problems of tranquilizing a wolf in subzero weather. Sage could have been in the trap for several days; he had surely suffered only frostbite at best, and he could lose his foot or his life at worst. By now, it was late afternoon, sunlight was fading, and we had been unable to contact wildlife officials in British Columbia to request permission to free Sage—presuming he would still be alive by the time we made the long trip around to where I had last seen him.

We just kept driving and dialing our clunky mobile phone (it was pre-cell phone days), until finally we got through to a biologist in Cranbrook—but he could not give us permission to free Sage. He said we needed to call the top wildlife administrator in the provincial capital, Victoria, and we could not release the wolf without proper permission. So I phoned the director of the British Columbia Wildlife Branch at his Victoria home at 9 PM. I could hear in the background that he had a houseful of people partying, and I apologized for the intrusion. I explained who we were, described our situation with Sage, and asked permission to release the wolf from the trap. The director gave his permission, requested a write-up of our release efforts, and wished us a Happy New Year. I thanked him heartily, and we headed for the Wigwam.

We drove from an icy highway, to a reasonable gravel road, to a nasty gravel road, to our turnoff into the Wigwam drainage, where we could drive no farther. It was 11 PM. Then we saw fresh tracks and realized that Henry had arrived and gone into his camp since my morning flight. We quickly unloaded the snowmobiles and began the long ride up the Wigwam drainage. The propane lights were on when we arrived at Henry's camp. We knocked on the door, and we were greeted by Henry's wife with a raised glass and a "Happy New Year! Who are you?" We introduced ourselves, told her of our mission, and peeked behind her, expecting to see Sage's bloody carcass in the hallway. As it turned out, they had arrived after dark so Henry hadn't checked his traps yet. They had been liberally celebrating the New Year with libations while warming up the subzero cabin. I asked to speak to Henry. The wife looked at me and said solemnly, "I don't think I'll wake him for this."

Mike and I left the cabin and organized our gear, sleeping bag, headlamps, and capture equipment. We opted to leave the snow-shoes at the snowmobile, figuring they would be more nuisance than help. The temperature was seriously below zero as we headed to the spot on the river where I had last seen Sage. The full moon lit up the river landscape so brightly that we didn't need to turn on our headlamps. The cold, dry snow squeaked beneath our Pac boots; the air was dead still, but Sage was, hopefully, alive and waiting.

We walked to the spot on the gravel bar where Dave and I had first seen Sage that morning. The drama was evident in the moon-light. The willows were flattened and chewed down, signs that Sage had been trying to extricate himself from the cold steel clamped on his foot. We found the bait carcass at the capture site.

"Looks like Henry used the back half of a moose for bait," I said as I gazed down at the black haunch in front of us.

Mike looked a little more closely and said wryly, "Hmmm, looks like that moose was shod."

Sure enough, Henry had used a hindquarter of one of his horses for bait, and the moonlight glinted off the steel shoe.

We followed the furrow made by the struggling wolf to the river's edge, where it disappeared into the water. Mike and I slipped into our chest waders and crossed the river, marveling at the slush ice floating past our legs. We changed back into our Pac boots on the other side, and it was only a matter of minutes until our discarded waders were frozen into a stiff, rubbery pile. We thought about Sage—no circulation in the foot clamped tightly in the trap, several days of subzero temperatures, his restricted mobility, and his wan-ing strength. We continued along Sage's wide drag trail, leaving the river bottom, and we entered a dark spruce forest. Here we had to turn on our headlamps. Sage had dragged the anchor log up a steep slope for a hundred yards; it was steep enough that Mike and

I had to pull ourselves up the hill with the aid of brush and sapling trunks. My headlamp lit up the shine of a steel chain, stretched tightly over a large deadfall lying across the ascending trail in front of us. All was silent as we stared at that chain. I slowly climbed uphill and peered over the top of a log—a massive, gray, frosty face rose up and stared at me, the face of a much wiser wolf than the one I had seen howling by my cabin. I slowly backed down toward Mike and whispered, "He's alive."

WE PULLED OUT all the equipment we would need to drug and take care of Sage, defrost his foot, keep him warm, and replace his radio collar. It was so cold that our liquid drugs had frozen in their vials, so we had to briefly thaw them in our armpits. A few minutes after we injected the exhausted wolf, his head dropped to the snow and we began our work. My watch read 1:30 AM as we worked together to remove the trap from Sage's frozen foot. From the ankle down, his paw was cold and hard as stone. For more than an hour, we took turns wrapping our mitten-warmed bare hands around Sage's frozen paw—being careful not to rub it, to minimize further tissue damage. We pressed our warm flesh against his wood-like paw, awaiting signs of thawing. The paw eventually began to soften. When the toes were pliable, we gently worked our fingers between the wolf's toes while wrapping another warm hand around the paw. The foot eventually became flexible and warm. We wrapped him snuggly in my sleeping bag, with only his affected foot and snout poking out. We replaced his old radio collar with a new one, checked his vital signs, injected antibiotics, filled out the capture data form, and continued the wolf-care procedures.

We waited for Sage to wake up and walk away. Two hours had passed since we had drugged him and yet the big wolf did not move. We had used a light dose of a fast-acting drug—ketamine—and he should have been up and gone in less than an hour. Mike and I stood there in the moonlight, stomping our chilled feet and rubbing our arms and legs to keep warm. Another hour went by and still the wolf lay there, his massive muzzle sticking out of the sleeping bag, breathing deeply and slowly. As the hours passed without movement from Sage, bleak thoughts began to creep into our heads. Had he become so exhausted from his epic ordeal that he would die despite our efforts? Had he dislocated a hip dragging the heavy log? Had he undergone organ failure? I thought to myself how terrible it would be if we freed him and he died, despite our ministrations.

At 4 AM, I couldn't stand it anymore. I grabbed a stick and approached the still form in the sleeping bag. I poked Sage and he jumped up, the sleeping bag sliding off into the snow. He looked even more surprised than we did; he shook himself off, stared at us for several seconds, and limped off into the shadows of the forest. Apparently, he had been awake for quite some time, but comfortably tucked inside the warm cocoon, he had been unaware that the trap that had held him tight for many days was gone. Mike and I looked at each other, shared a hug, and decided it was by far the best New Year's that we had ever experienced. We packed up our gear, retraced our steps across the wintry landscape, left the mangled wolf trap on Henry's doorstep, started our snowmobiles, and headed back up the trail to our pickup. It was 6 AM by the time we reached our truck.

We watched Sage intently from the air during our next several tracking flights and saw that he limped badly on the damaged hind leg for several months. But eventually he regained a more natural gait, the foot remained attached to the leg, and it seemed as if he had survived his ordeal in fair shape. He responded to Henry's trapping by denning a few miles from Henry's camp in the spring of 1988, leaving scats and tracks as a reminder that the big gray was alive and well. Sage and his mate produced a litter of six pups, and Sage's tenacious genes were passed on once again.

But Henry may have written the final chapter in Sage's life— Sage's radio collar stopped transmitting during the fall hunting season. With no other radio-collared wolves in the inaccessible Wigwam valley, it was difficult to monitor the pack. By the spring of 1989, evidence of the pack ceased, with no sign of wolves at the traditional denning area or anywhere else in the Wigwam area. The Wigwam Pack had disappeared as quietly as the melting snowpack. In August 1989, we captured and radio-collared a handsome, young gray male (8910) in the Wigwam valley who was joined by an uncollared black female the following winter. We never knew whether or not those wolves were remnants of the Wigwam Pack, but the cycle began anew.

SAGE'S STORY UNFOLDED through the help of radio telemetry, many miles of snow tracking, and the passion of a handful of young wolf researchers. The most significant challenge that wolves in the North Fork faced then are no different from the challenge wolves nearly everywhere face today: coexisting with humans on a human-dominated landscape. As a biologist, I relay Sage's story at the risk of appearing sentimental, but as the famed evolutionary biologist Stephen Jay Gould so eloquently put it, "We cannot win this battle to save species and environments without forging an emotional bond between ourselves and nature as well—for we will not fight to save what we do not love"—and wolves have definitely snagged my heart.

We don't get to really know the wild wolves we study; we have to be content trying to comprehend the ecology of these complex creatures via brief glimpses into their lives. Sage may typify the fate of a wild wolf anywhere in the world: they are born, learn life's lessons, disperse, try to establish a territory, leave progeny if they are lucky, avoid humans, and then they die—most likely due to human causes. During the years of our research in the North Fork, from 1979 to 1997, wolves were a federally protected endangered species in Montana and a regulated trophy animal in British Columbia and Alberta, yet we found that 85 percent of the wolf deaths in our study area were human-caused. Despite this trend, wolf recovery in the Lower 48 has been one of the most successful endangered species recovery stories in the U.S. since the Endangered Species Act was passed in 1973.

7

...

CATCH ME
IF YOU CAN

I N ALL MY years working for the Wolf Ecology Project, winter
was the season of wolf tracking and observation from the air,
then skiing on wolf trails and learning as much as possible about
wolf ecology by reading their stories in the snow. Summer days,
in contrast, were spent trapping and radio-collaring wolves, which
meant many days of boredom in warm, mosquito-infested forests
and marshes, punctuated by a couple of hours of excitement.

MY TRAP WAS gone, a pungent scent hung in the air, a pawprint
the size of my open hand was faintly visible in the floury dirt, and
signs of a struggle marked the roadside where my trap used to
be. I could hear an animal breaking off saplings in the lodgepole
thicket, but I couldn't see it. I took a few more steps toward the
commotion. Suddenly, all went quiet. I crept in closer holding a
custom-made jab stick I'd whittled from a four-foot willow sapling,
with a syringe loaded with tranquilizer at the far end. Suddenly, an
enormous black wolf leapt up from the underbrush—the formida-
ble, 112-pound dominant male of the Sage Creek Pack.

This wolf had probably not been afraid of anything in a very
long time, but now he struggled to get away from me, lunging to
the far end of the trap's eight-foot chain. I calmly approached him,
and he realized his number was up. He turned and faced me, his

amber eyes locking with mine. Then he stopped struggling, gently lowered his head, laid his ears down, and slowly began wagging his long tail in low, sweeping arcs, in apologetic submission. I felt terrible having reduced this magnificent animal to begging for his life. I poked the drugs into his rump, backed off, and waited for the tranquilizer to take effect. Andrea Blakesley—my WEP technician—and I repeatedly took his vitals (pulse, respirations, rectal temperature) and after determining he was stabilized, we drew blood samples, weighed and measured him, attached a radio collar, and named him Ace, because he was definitely number one.

Matching wits with wolves was an exciting part of my professional life. When I was learning how to think like a wolf and trying to lure the cunning animal to step on an Oreo cookie–sized trigger, I made a lot of mistakes. In my early North Fork years, I captured nontarget species, including coyotes, lynx, bobcats, mountain lions, and black bears. Fortunately for me and the resident wildlife, I learned how to place traps and lures more selectively so that I began to catch mostly wolves and a lot fewer of the other species. But before I improved my techniques, I had several memorable moments, including the day I trapped my one and only grizzly bear.

I was up in the Dutch Creek drainage in Glacier National Park. My trap was gone, and I could hear a big animal rustling around in the brush—but I couldn't see it. I was presuming, excitedly, I had a wolf. Or I was, until I saw a tan-and-silver dished face peek through the willows at me. Oh dear God, I had accidentally trapped an adult grizzly bear. I got on my park radio and contacted headquarters. As my call was being broadcast throughout the entire park repeater system, I was trying to let them know discreetly that I needed some ranger assistance as soon as possible. I did not want to blurt out that I had a grizzly in a leghold trap, create a lot of alarm, and have a squad of armed rangers flocking in.

The Polebridge rangers Scott Emmerich and Regi Altop, whom I'd worked with for years, soon arrived on the scene. They were both trained in bear management and dangerous situations. They worked together to get close enough to the hidden grizzly to shoot a tranquilizer dart into it without endangering themselves or the bear. They needed the right angle, but the bear—a female, as it turned out—uncooperatively stayed quietly hidden in the brush. Finally, she turned to look away, and Regi fired a dart that hit her perfectly in the shoulder. A few minutes later she was asleep, and it was safe to approach. We all swallowed hard when we saw that she was caught by only two claws. If she had lunged hard once, she would have popped free of the trap and who knows what might have happened. I removed the trap from her claws, and we sat in the protection of a distant vehicle to make sure she came out of the drugs all right. She stumbled away safely after an hour.

A YEAR LATER, Dutch Creek delivered another of her secretive creatures to my trapline. The thrashing in the brush told me I had something caught in my trap. I approached cautiously and saw a small mountain lion face peeking out of the willows. I had never caught or even handled a lion before. My friend and colleague Wendy Arjo was helping me check traps, and we were thrilled to be able to get our hands on a lion. Wendy and I were conducting our PhD fieldwork simultaneously and were helping each other capture and collar wolves and coyotes for our research. Wendy was bright, bubbly, and brave, and I admired her spunk; we grew to be close friends.

We couldn't see the lion's body, but judging by its head size I guessed it was a kitten, weighing about thirty pounds. I loaded up my syringe with a small amount of tranquilizer and walked up confidently to the hidden lion. When I got within twenty feet, it launched out of the willows at me—six feet of tawny, muscular power, plus another three feet of tail. Definitely not a kitten. It then dawned on me that wolves have huge heads for grabbing prey and relatively smaller, lean, leggy bodies; whereas mountain lions have huge, powerful bodies and legs for grabbing prey and relatively small heads for delivering a lethal bite. This adult female lion's petite head was attached to a ninety-five-pound body that was all muscle. Regroup. I told Wendy I'd go for it and underdose the big cat so it could walk away sooner. But first we'd put a wolf radio collar on her for a mountain lion study being run by Toni Ruth of the Hornocker Wildlife Institute in Moscow, Idaho.

The lion seemed to know that I was the business end of this pair of humans and followed only me with her gaze and growling, no matter how much Wendy clapped, hollered, or stomped her feet. "Wendy, distract her with something bigger," I said. Wendy picked up a thick, long stick and whopped that cat on the nose with a resounding thunk. "You've got guts, girl!" I thought. With that indignity, the lion whipped around and, in a flash, swiped at Wendy with a large, sabered front paw. In that moment, its rear end was exposed, so I lunged forward and poked the lion in the rump with my jab stick and quickly stepped backward. Done.

I hoped I had enough drugs loaded in the jab stick to put the lion under, since I had grossly underestimated her weight. As it turned out I did, but barely. She slowly lowered to the ground, but her head never went fully down, and she looked around unfocused and angry while her tail twitched and lashed around. We were careful to avoid her mouth while fitting her with her new radio collar. I felt her thick, strong forelegs and marveled at what an incredibly powerful killing machine she was. How many deer and elk had this gal stealthily stalked and killed in her lifetime? She was one of the most magnificent animals I had ever touched. We had no time to get her weight, blood sample, measurements, or photos because she got up and walked away a few minutes after we attached the collar.

ON ANOTHER OCCASION, in the Sage Creek drainage of British Columbia, I was checking my traps alone and discovered that one trap I had set at the base of a tall lodgepole pine was gone, with no drag trail, tracks, or sign of struggle leading away from the site. I was wondering what had happened when I heard branches break overhead. I looked up and saw a small, brown-colored black bear perched near the top of the tree with my trap on its foot. The chain and anchor hook were jammed into the branches five feet above it. The yearling cub must have climbed nearly to the top, where it hung up the anchor, and then descended down the tree until the chain tightened.

I looked around anxiously for an angry momma bear to come charging in any moment, but she didn't materialize. I went to my truck and turned my dog Stony loose so he could alert me if he saw or smelled the mother bear. I loaded up the syringe on the end of my jab stick with tranquilizing drugs and began to climb up the tree toward the cub, who now began bawling in dismay. Do your job, Stony. Good boy. I stopped a few feet below the bear and gave it a gentle poke in the rump with the drugs. As the tranquilizer began to take effect, I had to climb above the woozy bear to free the tangled anchor hook—before the bear went limp, lost its grip on the tree trunk, and fell out of the tree. I tapped the bear on the snout with the jab stick and it huffed and tried to swat me, so I backed down a few feet. Then its claws began to loosen their grip on the tree, so I tapped it again and it didn't respond. I quickly scrambled over the bear, reached up and freed the anchor hook, shinnied back down to the bear, and tucked it under my left arm. Using my right arm, I slowly climbed down the remaining twenty-five feet to the ground, limp baby bear safely in my arms. Stony badly wanted a piece of that bear, so I wrangled Stony back into the truck, much to his disappointment and loud protestations. I removed the trap from the bear's foot, laid it on its sternum in the shade, checked it over to make sure it was all right, and left it to sleep off its very bad day.

FLYING AND LOCATING radio-collared wolves was my favorite part of the WEP fieldwork. I loved soaring over incredibly wild and rugged country listening for the pings of the radio collars through our headsets. Each wolf had its own unique frequency so they could easily be told apart. We quickly went in on mortality signals so we could investigate the cause of death before the wolf was consumed by bears and other scavengers. From aerial surveys, we learned about pack size, preferred habitat and travel routes, chase scenarios, and prey characteristics. When I was following wolf trails on skis, I read their stories in the snow—on what they did when they encountered the scent of other wolves, coyotes, deer, elk, and bears, where they slept, how they approached an unusual smell, and how they surveyed their landscape. After skiing a few hundred miles along wolf trails, I learned how far they could detect a scent, what smells were irresistible, which were ignored, and those that were feared. A squirrel track barely deserved a sniff. A coyote or lion track elicited some interest and maybe a little scent mark over the tracks to show who was boss. But the track of a foreign wolf—oh baby, game on!—as the wolf stuck its nose in several prints and followed its competitor's path. Along the way, I would find scent posts left as olfactory sticky notes, deep scratches tearing up the snow or bare ground as visual clues, and maybe a good roll in the white stuff that left wolfy angels. The WEP field crew would collect concentrated snowy urinations in ziplock bags and bring them back to me. I thawed the frozen pee in glass jars and used it for summer wolf captures. Scats were also bagged, labeled, and collected for food content analysis, as well as trapping scent.

Tracking wolves in winter taught me more about how to catch them in summer than the plentiful advice I received from old-time wolfers and modern trappers. And the more years I trapped these smart and noble animals, the more regret I felt at causing them trauma. I apologized to them before I injected a sedative and freed them from the trap to fit them with their new radio collar. When you collar one animal in a pack, you gain information on that wolf as well as the other pack members, and I firmly believed that fitting a wolf with a radio collar and gaining data from it was worth the wolf's temporary capture distress so that we could learn more about the species for their conservation.

I was thrilled with every wolf capture and was dismayed by every one that I missed. The missed wolves were my best teachers when it came to improving my trapping techniques—and the misses far outweighed the successes. A trap night is one trap out for one night. My traplines often had a dozen traps out, so a week would be twelve traps times seven nights, equaling eighty-four trap nights. As I was sharpening my skills, I took pride in catching a wolf in less than one hundred trap nights. For every wolf caught, there were dozens of days of nothing. Sometimes, it took two hundred or more trap nights to catch one wolf in a new area of low wolf density.

A WOLF'S SENSE of smell is incredibly keen. Their bodies have evolved to be mass emitters and detectors of scents and pheromones, and their survival may depend on being able to assess these scents quickly. Wolves produce scent in their urine, feces, saliva, anal glands, foot glands, tail glands, and vaginas. A wolf's sense of smell is approximately one hundred times better than a human's and about the same as a domestic dog's. Wolves can detect prey up to one and a half miles upwind. Trying to outsmart a wolf is the ultimate challenge.

The grizzly bear trappers used mountains of rotting meat and putrid blood, often dragging a deer hindquarter down the road behind their truck. They called it "walking the dog," and this method lured in bears, and occasionally wolves, from miles away to their trap sites. This is how they accidentally captured a few wolves in their heavy-duty cable bear snares. A rotting mountain of meat is nearly irresistible to both bears and wolves, but I was using a leghold trap that wouldn't securely hold a bear. If I accidentally caught one or, worse yet, caught a cub while a free and angry mother bear was nearby, it could get me killed—and so I avoided over-baiting. Instead, I used a few drops of a lure or a kidney-bean-sized piece of bait to catch wolves, while—usually—avoiding catching bears.

The key to success was that I had to try to eliminate all human scent. To minimize scent of the traps, I boiled them with alder bark and branches in a fifty-five-gallon drum outdoors until the tannins turned the water black and darkened the steel traps. Using a long stick, I lifted the steaming traps out of the boiling witches' brew and hung them on the limbs of spruce and fir trees by my cabin. They would weather there for several weeks before I would use them, letting wind and rain wash away any residual odors. Future handling of the traps was always done carefully with gloves that had been buried in dirt to be scentless. I made a special box filled with pine boughs in the back of my truck that carried traps and trapping gear. I sincerely believed that all of this might fool a few wolves into stepping on my trap so I could radio-collar them and study them.

Placement of the trap is also key—location, location, location. You can make a fabulous, well-camouflaged set with an irresistible lure, but if it's not where wolves travel you aren't going to catch one. When I was out hiking, skiing, or driving back roads, I began to match up the landscape with where I thought a wolf would travel and what would snag its attention. With wolves, less is more. Less disturbance, less scent, less sign of my presence, and just a hint of lure to bring them in close for a more thorough exploration, without drawing in nontarget species.

I took pride in setting and concealing a trap in ten minutes, when everything went right. That happened when the trap didn't accidentally trigger, launching dirt shrapnel into my face and eyes, or when I didn't hit a big root in the middle of the hole I was digging, forcing me to move to a new spot to start over. To make myself be extra careful yet quick, I imagined the trapping site covered in a paper-thin layer of fine white flour, where every footprint and handprint was wet and sticky—and would show up contrasting sharply as dark dirt through white powder. I saw this in my mind, but a wolf sees these footprints through its nose—and every step and pause marked with my smell would scream danger. Wolves drink in the landscape through their all-powerful nose, detecting every molecule of foreign scent in their familiar home. Studies have been done that have measured the thousands of molecules that come off a human per minute. Wolf survival has been honed to detect these markers, be they from humans, prey, or competing predators.

I conducted the majority of my research trapping during the non-freezing months of May through October to minimize the possibility of freezing toes, since the trap closes tightly and might cut off circulation. However, trapping during summer months can expose animals to heat stress, so on warmer days I checked my traps two to three times per day to make sure a wolf wasn't overheating. I placed most of my traps in shady areas along the edges of roads or trails, because that's where wolves frequently travel. Additionally, this allowed me to check my traps from a pickup, driving slowly past each trap without leaving any of my scent on the ground.

KURT ALUZAS, a volunteer during the summer and fall of 1988, always seemed to be with me during some of the more exciting capture events. On June 8, 1988, we recaptured a black radio-collared female wolf, 8756, on the Lower Sage Creek Road just north of Glacier National Park. I had tranquilized her with the jab stick, and we had mostly finished up processing her: radio collar fitted, blood sample drawn, vitals monitored, measurements taken, etc. All that was left to do was to sit with her while the drugs wore off to ensure she was all right, pack up our capture kit and field gear, and leave the site when she got up safely and woozily walked off. She was beginning to wake up with the usual head bobs, so I knew it wouldn't be long until she was mobile.

As we were quietly packing up our gear, we heard a stick snap.

Kurt's eyes got big and he said, "There's a bear."

I looked where he was staring and said, "That's not just a bear; that's a grizzly bear."

The big silvertip had apparently smelled the anal-gland fear scent given off by 8756 in the trap—and perhaps smelled the blood from the trap puncture on her foot as well.

I stood up, slapped the clipboard, and hollered, "Hey, bear!"

The bear quickly turned around and disappeared into the thick lodgepole forest. The shouting made the wolf lift her head, and she tried to roll up onto her chest but couldn't quite manage it yet. I told Kurt to go get his Jeep, which was parked a hundred yards away on the road. Kurt quickly took off and headed toward the road. I could hear him trying to start his Jeep, which had a carburetor problem. While Kurt was cranking away, trying to turn over the engine, I looked up to see that the grizzly had returned and this time, it was a little closer. It stood up and slowly swung its head back and forth, sniffing the air only thirty yards away from me.

"Hurry, Kurt. Come over here!" I bellowed.

Disturbed by my yelling, the grizzly dropped back down on all fours and disappeared again. Finally, the engine caught and Kurt moved the Jeep closer to me, cleverly wedging an axe between the front edge of the seat and the gas pedal so the engine revved at a high whine and wouldn't die. Then he walked briskly into the forest, wide-open eyes scanning everywhere at once, to where I was trying to gather up the wolf.

We looked up and saw that the bear was back again, now only twenty yards away, and fixated on the semi-tranquilized wolf. It's strange how your mind intensely focuses and channels your thoughts in time of crisis. All I could think was that I had put this wolf in this dangerous predicament, and the grizzly wanted to kill and consume this competitor; the bear was focused on the wolf and was not going to kill us humans—although it might perceive us as thieves stealing his meal and charge us to claim the wolf.

I slapped the clipboard again and yelled, "Get outta here, bear!"

I grabbed the front end of the wolf while Kurt grabbed the hind end, and we rapidly manhandled the partially drugged and now struggling wolf out to the running Jeep. I only looked forward toward the Jeep and had no idea where the bear was. I opened the back door, and we stuffed the wolf in the back seat, headfirst, up against my big dog Max—who wasn't happy about this dopey, wild creature shoved against his feet that was trying to bite him but was unable to control its head movements due to the drugs.

We jumped into the Jeep and screamed down the road in first gear with the rpms nearly pegging out on the tachometer, because the wolf's rear end had fallen between the seats and was jammed against the stick shift, holding it in the forward position. "Pull in here," I ordered as we came to a small, two-track road, where we slid to a halt and quickly unloaded the awakening wolf. We made her comfortable in a shady spot under a large Douglas fir and backed away about seventy-five yards to where we could sit down to watch her come out of her tranquilized state unmolested. I handed Kurt a can of soda and he just looked at me.

"Well done," I said and smiled.

I WAS CHECKING my traps in early December, thinking that with the winter weather settling in I should shut down capture efforts for the season. But just a few more days and maybe I would catch a wolf. Sliding sideways in my pickup, I negotiated an icy hill in four-wheel drive just before Cabin Creek, trying not to go into the ditch. As I came around the curve, I saw a dark animal jumping in my trap. Wolverine! I had never seen a wolverine before but there was no mistaking it—that churning brown back with the golden side stripes was Ms. Badass herself.

I got out of the pickup, shaking with excitement. I calculated the dosage required to tranquilize a thirty-pound animal and loaded up the syringe. I mounted the syringe on the end of a four-foot jab stick and cautiously approached the trapped wolverine. When I got within thirty feet, she launched herself at me, hit the end of the chain, and flipped over. Oh. My. God.

Wolves, coyotes, and most other animals I had approached with my magic jab stick cowered or turned away trying to escape. Not this old wolverine queen, no way. I warily approached her, my heart pounding, as she kept launching and lunging at me. She was an angry tornado on the end of an eight-foot chain. Her tactics of intimidation were working to erode my confidence. I didn't want to hurt her. What if I couldn't get the drugs into her? What if she tore off her foot? What if she pulled free and ripped my throat out? Fearsome old trappers' tales about a wolverine being ferocious enough to fend off a pack of wolves at a kill, or to dispatch a polar bear single-handedly, were coming back to me all too clearly.

When I got within poking distance, she hurled herself at me through the air again. As she reached the end of the chain she flipped over, and I lunged forward and poked her solidly in the rump with the syringe. I backed away and waited quietly, relieved at finally getting the job done. Time passed slowly as the wolverine and I stared each other down. The sedative usually takes effect in three minutes. But ten minutes turned into twenty, and yet the wolverine was still tearing things up, growling, and giving no sign of having any drugs in her system.

I realized that I needed to repeat the fencing scenario with the jab stick. I loaded up another smaller dose of drugs, quite sure I had gotten some of the drug into her on the first attempt and not wanting to overdose her. We repeated this wild launch-and-lunge dance until I got close enough to poke her in midair as she flipped over while trying to kill me. She began to wobble and head-bob in two minutes, and in another two minutes she was completely out of it. I approached her and lightly tapped her with my jab stick. She did not respond, so it was safe to begin working on her.

I removed the trap from her right front paw and checked her over carefully. This old gal could tell some amazing tales, judging by the long scar that ran out of the corner of her right eye, the milky film over her left eye, and her yellowed, worn-down teeth. Yet she was in good shape except for her trapped right paw, which I had unintentionally mangled in the capture event and her subsequent struggles. My feeling of excited awe was replaced with remorse for having hurt this creature, the most amazing animal I had ever seen. On top of that, she stayed drugged longer than I thought normal, and then I realized that she had likely received the full drug dose on the first injection but was so pumped up that her adrenaline overrode the drug's effect. The second injection had sent her over that threshold and knocked her out completely.

I checked her vitals, treated her foot injury, and made her comfortable in a snowy, insulated bed. Then I sat back fifty yards from her to watch over her until she came out of the drugs and safely sauntered off. She took more than an hour to begin stirring, as I sat shivering in the snowbank with binoculars staring at her, willing her to wake up. She slowly became aware of her surroundings in her fading psychedelic trip. The second she saw my form crouched in the snow, she locked onto me like a lethargic heat-seeking missile. And then she began to drag herself through the snow toward me.

I moved off to the side a few feet, and she compensated and changed her direction—and kept crawling toward me. A little unnerved, I backed toward the truck as she struggled to get on her feet and then headed in my direction. She was well on her way to recovery, and I was on my way out of there. I came back in a couple of hours to check on her, and thankfully she was gone. But her trail in the snow clearly showed that she had gone directly to where I had been sitting and pissed over my smell with her skunky urine before disappearing into the spruce trees.

The next day, I was out checking my trapline by myself and I had yet another flat tire on the ice-rutted road two miles from home. Damn those worn-out tires. On the icy road, I jacked up the back end of the pickup with the bumper jack high enough to remove the flat tire and replace it with the inflated spare. It was below zero, so I wore my wool mittens with leather chopper over-mitts while changing the tire. I pulled the flat tire off the lug bolts. As I was pushing the inflated spare onto the lug bolts, with my right hand on top of the tire and left hand under it, the bumper jack suddenly shot out from under the truck. The truck crashed down onto my right hand before bouncing up briefly from the inflated spare. I instantly pulled my right arm back and fell over backward as the truck crashed to the road and the tire fell off to the side. I looked at my right hand, which had been crushed but was effectively splinted inside the big mitt. I had some time to contemplate what I was going to do now that I couldn't change the tire to get myself out of there. The truck had nearly trapped me by my right hand. And then I remembered yesterday's wolverine capture . . . by the right front paw. The curse of the wolverine.

I walked three miles back to Moose City and asked a friend, Rick, to help me. He took one look at my face and jumped into action to rush me to the emergency room, only two hours away. Wait, not quite so fast—we had to pull all my traps before we went to the hospital so that no animals would be trapped while I was in town being treated for my broken hand. I directed Rick, in his truck, to each trap site, and he sprung them with a shovel and pulled them out of the snow. Locating and closing the trapline took a couple of extra hours, but I was reassured that no animal would be hurt due to my carelessness. Then I relaxed and let my friend and the medical staff take care of me.

WHILE I WAS working with the Wolf Ecology Project, I was also completing my PhD at the University of Montana. My dissertation covered several aspects of wolf ecology, but my favorite topic was wolf dispersal because it was the key to wolf recovery: How, when, and who would leave their natal pack and migrate to a new area to find another pack or a mate? How long did it take wolves to make the final break from home? How far did dispersers travel, in which direction, and why did some die and others succeed?

One February day, my pilot Dave and I were doing a dispersal flight, searching high and far away from the North Fork, hoping to find some of our wolves that had gone walkabout. We were flying at ten thousand feet for what seemed like hours, listening to the static of the receiver drone on in our headsets as it flipped between the frequencies that I had programmed into it for our wandering wolves. Dave, with eagle eyes and sensitive hearing, said, "Do you hear that?" And then I heard a faint, intermittent beep. I locked onto that frequency, and we kept flying as the signal grew stronger in the direction of the wolf. We homed in on that signal and began dropping down until we could see five wolves romping in a snowy mountain meadow.

Eventually we could see the radio collar on one of the females missing from the Kintla Pack. She was the center of attraction for a very large, attentive male, while the other pack members milled around excitedly. Wow. It was Valentine's Day, the peak of breeding season for wolves, and our Glacier wolf was being courted by a wolf from Alberta. I was excited to think that in sixty-two days, approximately Tax Day in the U.S., she would pay her dues and give birth to a litter in her new Canadian pack. Yes!

I smiled, turned to Dave, and said, "Where are we?"

"Beats me," he replied.

This was before the days of GPS and Google Maps, so our locations and reckoning were based on experience and paper maps. But we were so far north of Glacier Park that we had no idea where we were. Dave said, "I'll fly, you read," and I was a little puzzled until he banked the plane hard to the left and dropped down so we were about twenty feet above a closed, snowed-in highway. I saw some green exit signs coming up. As we flew past the first one, I quickly jotted down "Fitzsimmons Creek" followed by "Strawberry Creek." I gave Dave the thumbs-up, he pulled up on the controls, and we climbed up into the sky and headed home.

The next day I drove into Fernie, British Columbia, and bought several paper topographical maps that contained the home range of our new pack in the Kananaskis area. We got the exact location of our international lovers in a wild place where wolves would prosper. I called Canadian wolf researcher Paul Paquet and the wolf biologists in the Kananaskis area and gave them this wolf's frequency so they could continue to track this pack. It was beyond our limited budget to fly this far, and the Kananaskis biologists were thrilled to have a radio-collared wolf in their pack so they could track them.

OVER THE YEARS, we followed Flathead radio-collared wolves as they dispersed in all compass directions, like dandelion fuzz on a windy day. Wolves set up territories in the Ninemile drainage near Missoula; the Rocky Mountain Front near Choteau and Cut Bank, Montana; Kelly Creek in Idaho; and Waterton Lakes National Park and Banff National Park in Alberta, Canada. Throughout the 1980s, wolves dispersed south from Canada into Montana, followed by Montana wolves dispersing north into British Columbia and Alberta to backfill some low wolf density areas. Wolves that the WEP was studying in the Flathead eventually became the source that would recolonize many formerly wolfless areas.

It was thrilling to be a part of the unfolding story of how wolves were recolonizing their former home ranges, all by wolf paw power. I worked with Paul Paquet and Parks Canada to catch and radio-collar wolves in Banff, Yoho, and Kootenay National Parks in British Columbia and Alberta. Through this international effort, we learned that wolves disperse hundreds of miles across provincial, state, and international borders. But until we had radio collars to track them, we didn't know it.

Twenty years into the future, things would change yet again. The arrival of GPS radio collars allowed researchers to keep track of their animals multiple times every day, monitoring their dispersals and the migration corridors that are critical to wolves, grizzlies, and other wildlife to help keep their populations healthy and viable. Noninvasive genetic analyses and analyses of stable isotopes of carbon and nitrogen would also arrive on the scene as tools to collect information on carnivore dispersal, reproduction, food resources, and so much more. But not quite yet.

8

...

PHYLLIS

WEP FUNDING DRIED UP from 1982 until 1985, but I stayed on at Moose City as caretaker without the WEP crew. I sold my artwork, followed wolf tracks on skis, checked out wolf reports from my neighbors, and searched for wolves as my gas money allowed. I couldn't help myself; I was hooked. By the spring of 1985, after a three-year hiatus, the Wolf Ecology Project once again had funding. In the early years before funding was temporarily suspended, Mike Fairchild, numerous volunteers, and I had logged thousands of miles trying to capture wolves and scores of miles on skis following wolf tracks. But wolves were scarce and luck was with the wolves. Now, with funding restored and radio collars in hand, we were all set to try again.

On May 18, 1985, I came home from a long day in the field and found a note inside my Moose City cabin. It was from Canadian grizzly bear researcher Bruce McLellan saying he had accidentally captured a lactating female wolf in his bear snare, a few miles north of Glacier National Park. He had come down to get me, but since I was absent, he had taken a radio collar off my kitchen table to put it on the wolf for me. I was ecstatic and grateful.

To get the details, later that day I hiked up to Bruce's river crossing and canoed across the North Fork to find him at his cabin. He told me that the wolf was in good shape, although "a bit ribby" due to the energy demands of nursing a litter of pups. Pups! Bruce estimated that she was about seventy-five to eighty pounds, and maybe three years old based on tooth wear. Her outstanding feature was her creamy, nearly white fur, which made her easily identifiable with or without a collar, because white was a rare color morph for wolves in the Rocky Mountains. I decided to call her Phyllis, named after Phyllis Forbes, our elk-hunting friend. Phyllis could possibly have been one of Kishinena's 1982 pups, but none of Kishinena's offspring had a white coat; it seemed more likely that she had wandered down from the forested mountains of British Columbia or Alberta.

In southern Canada and northern Montana, a white coat is usually a badge of honor worn by a few old and lucky wolves. This is in contrast to Arctic wolves, which are normally white. Wolf fur, like human hair, turns white with age or, possibly, trauma. But this mama wolf was neither old nor injured; she just had an uncommon gene for the white coat color. In the old Wild West, many of the last lone wolves were described as white ghosts: alleged savage livestock killers, highly cunning hunters, and supremely difficult animals to catch in a trap. These last renegades were indeed smart and simply out-survived their less wary brethren, becoming old and solitary white wolves—often missing parts of feet or legs due to escaping from traps. Those injuries made the crippled wolves much more likely to catch a slow Hereford than a fleet-footed deer and, ironically, trapping may well have turned non-depredating wolves into livestock killers. At the same time, the surging white settler population and market hunters decimated game herds, replacing deer and elk with cattle and sheep, a fatal changeover for wolves across the West.

The day after Bruce collared Phyllis, I jumped into our mud-coated work truck and began searching with our telemetry antenna and receiver for the beeps of her collar. It was spring in the Flathead, the gravel roads were soft and potholed, and some were still blocked by snow. But there was a mother wolf out there somewhere near her den, suckling a litter of pups. I wanted to confirm from her signal that she was alive and that there was indeed a new wolf family in the neighborhood.

It was two weeks later during a tracking flight when we finally spotted Phyllis for the first time, outside her den nursing and playing with her seven black pups. She was just north of Glacier National Park in the territory of the Magic Pack, founded by Kishinena, the female who had started a whole new generation of wolves in the Flathead. After Kishinena's radio collar failed in 1982, we had no way to track her and she wasn't observed directly. She simply vanished. Whatever happened to Kishinena, Phyllis was now the dominant force to reckon with in the North Fork and the only breeding female in the area in 1985.

THE FOLLOWING YEAR, 1986, Phyllis decided her pack would benefit from a change in scenery, and she whelped five gray pups south of the international border, the first known wolf pack to den inside Glacier National Park—or in Montana for that matter—in fifty years. This was a huge news story, splashed across the U.S. media. Glacier National Park personnel were excited to add this pack to their wildlife database. The Magic Pack, as we continued to call this group, and its descendants have denned in Glacier National Park every year since.

Our fledgling Wolf Ecology Project also grew, thanks to increased funding from federal and private sources. Wolves were increasingly being recognized as rare, charismatic, and endangered. We had no shortage of enthusiastic college students who wanted to live in a remote cabin in the North Fork and track wolves for a few weeks or months, and the public was mostly buzzing with excitement for this new, wolfy phenomenon. And so began a dozen years of working with some truly amazing college students. These tireless volunteers were the backbone of our Wolf Ecology Project, working for no pay but all the scenery they could take in. It was a lot of hard work, dangerous at times, but also fun and profoundly rewarding.

The volunteers were game for whatever task was handed to them. In 1986, we remotely monitored Phyllis's radio collar locations from the North Fork Road every two hours around-the-clock during the spring denning season, leading to bleary-eyed volunteers spelling each other off, but good information gained. We documented den site attendance, as well as Phyllis's absence when she left the pups to join the other pack members on a hunting foray, nocturnal versus diurnal movement, how far she would roam from her den, and more. This was the first time this type of data had been collected in the western U.S. The North Fork was a place where grizzlies, black bears, cougars, wolverines, or even coyotes could enter a den and kill the pups. So who stayed behind to watch the pups? When did the mother wolf get relieved so she could go afield hunting with her packmates? Our research would help answer these questions.

STANDARD VHF COLLARS typically last for about four years, and our policy was to replace them before they failed while we could still track the wolves to trap them. Sadly, most wolf radio collars outlasted the life spans of the wolves wearing them. I spent many years trying to trap the elusive Phyllis, but she was a white phantom from the very beginning. I missed being in on her original capture by a matter of hours, and in the years that followed, I rarely saw her except for fleeting glimpses from the Cessna during our tracking flights. Despite our most cunning efforts, she never again got caught in a trap—and went on to teach her packmates and progeny how to avoid them.

I pulled out my best tricks to try to catch Phyllis. I boiled traps, gloves, and special boots in alder-bark water, then buried them in dirt and pine boughs for weeks, using only gloved, scent-free hands to touch them. Placing each trap, I worked quickly to minimize the scent molecules rolling off my body, while still making the trap blend perfectly into its surroundings. I took great pride in my trapping ability, and my traps were invisible to the human eye. I concocted a special foul-smelling lure to place behind the set—or sometimes in front of the set, for something different—or no scent at all. I kept switching it up in hopes of fooling her.

Despite my painstaking efforts, Phyllis walked past my traps and totally ignored them. But if she had packmates with her, she went to great extremes to share her trap wisdom. I could read the story in the mud: there would be tracks of several wolves convened around the trap site, and one smart wolf, presumedly Phyllis, working to extricate the trap. I pictured her family gathered around, getting the message—watch this, don't go near this foreign scent—seeing her ever-so-carefully pawing at what looked like an ordinary patch of dirt until a piece of metal was exposed. While her pack looked on, with heads cocked and tails drooping, Phyllis would cautiously and completely extricate the trap that I had so carefully bedded and concealed. She could pull it out from the hole, sometimes even flipping the trap upside down without springing it. Oh, that girl had a supernaturally delicate touch. And then, in smug satisfaction, she would defecate just to the side of the trap, proclaiming this turf as hers. She was uncatchable. To the best of my knowledge, she had never been caught in a wolf trap, yet she knew exactly what they were. So why did she mess with them at all? Phyllis drove me crazy—and I so admired her for that.

AFTER A FEW YEARS, something shook up the pack dynamics. A younger female—8653, whom we called Mojave—joined the Magic Pack in January 1987 and became the breeding female. Deposed, Phyllis left her pack to wander alone throughout the northern portion of her range in British Columbia. Sometimes she came by her old haunts in the darkness of winter nights and howled, a distinctive, breaking two-note song that would drop off mournfully. The first time we heard that howl at the Moose City cabins, we dropped what we were doing and ran outside with a receiver, which pinpointed the sound of Phyllis's frequency, coming directly across the frozen North Fork Flathead River. No other wolves answered her plaintive call, though it was nearly breeding season.

But love was in the air, and her pungent, bloody urinations in the snow advertised her fertile state. Somehow, she delayed or lengthened her conception date, which was highly unusual in wolves, denning one month later than in her previous denning years. She eventually mated and denned in Canada, half a mile north of her original 1985 den location.

Phyllis was a temptress and not to be outdone by a young upstart. Over the spring months, wolves drifted away from Mojave and the Magic Pack, who were denned twenty-five miles to the south, and joined Phyllis to help raise her third litter of pups in the Flathead. By midsummer, the two related packs were now approximately equal in size, with some members traveling between the two dens and provisioning both mothers and their pups.

Things changed when Sitka, one of Phyllis's helpers and probably her daughter, was illegally shot on September 18 in British Columbia. Then, on September 27, Canadian hunters shot one of Phyllis's pups. A month later, three more of her pups were shot and killed in Canada. At that point, all the wolves that had been with Phyllis returned south into the safety of Glacier National Park and rejoined the Magic Pack, seemingly understanding the danger of the British Columbia wolf hunting season.

This left Phyllis alone to raise her last remaining pup and find a safe haven north of the border. But within a month, that single pup disappeared, and Phyllis resumed her solitary roaming, first within her familiar territory and then to places far beyond in Alberta. We didn't hear her signal in the Flathead again until the beginning of 1988. Her radio collar was failing, and we heard its last beeps on February 28, 1988, far north of her traditional den and territory. After that, we lost track of her via radio signals, but for the next two years we continued to receive occasional reports of a white wolf wearing a black radio collar near Crowsnest Pass, B.C., and in Alberta along the Rocky Mountain Front. I wondered what she was looking for.

THE NEXT TIME I heard about Phyllis's whereabouts was from an Alberta game warden. He told me that a white wolf wearing a black radio collar had been spotted a few times in the forested hills of the Westcastle area, beginning in January 1990. Phyllis had apparently established a new territory along the foothills of the Rockies in Alberta and was often spotted near the Westcastle Minimum Security Forestry Camp, sometimes alone, sometimes with two semi-wild dogs.

Now entering her ninth year of life, quite elderly for a wild wolf, she became less wary and was seen more and more by camp staff, recreationists, and an outfitter and cougar hunter named Brent. Phyllis was spotted and killed by him on December 19, 1992. That's when the game warden called me to ask if I wanted the radio collar back. Yes and no—difficult question. Yes, I wanted the radio collar back to examine it and confirm from the frequency that it was Phyllis. No, because I didn't want to acknowledge that she was dead. She had been a famous wolf and her death sparked an outcry in headlines such as "Canadian Outfitter Slays Matriarch" from several Montana newspapers to the *Denver Post*, *New York Times*, and the British *Tribune*.

A few months later, I headed up to the Westcastle area near Pincher Creek, Alberta, with an environmental studies graduate student, Tommy Youngblood Peterson. He was writing his thesis on humans and their relationship to wolves. He had read about Phyllis's death and wanted to know more about it. I had come around to wanting to know more about Phyllis's death, so now that made two of us. Why would the wiliest of wolves, in her old age, become a regular visitor to a human camp and keep the company of dogs and humans? I was curious to meet the guy who shot her, to understand his motives. I wanted to close the loop on this ghost wolf who held a special place in my heart. Through hundreds of radio locations, she had given us glimpses into the secret lives of wolves, insights into denning behavior, and the survival of a species. Having birthed seventeen pups, she was truly the matriarch of the North Fork.

THE HIGHWAY WAS snow-packed and icy up over Marias Pass, along the windy Rocky Mountain Front and on to Pincher Creek in Alberta. As we rolled on through the intermittent black ice, high winds, and whiteouts accompanied by a big thermos of hot tea and a one-pound bag of M&M's, Tommy and I speculated on what we might discover. On the way to finding answers about Phyllis, we stayed overnight with Jean and Dave Sheppard in their homey log house in the foothills of the Rocky Mountains and were generously fed a huge dinner upon our arrival.

I met Jean and Dave at a wolf talk I had given in Pincher Creek the previous fall. Dave was the executive director of the Castle-Crown Wilderness Coalition, and he and Jean had been fighting a proposed winter resort development in the area. They were dedicated conservationists, and as retired teachers, activists, and longtime Pincher Creek residents, they knew everyone in the community. They had crossed paths with Brent at hearings about the proposed development, which Brent was also fighting. They all wanted to protect this excellent wildlife habitat, albeit for different reasons: Dave and Jean for the intrinsic value of wilderness and wildlife, and Brent to preserve the wilderness habitat and wildlife for his outfitting business and the big-game and predator hunts it offered.

THE NEXT DAY, Tommy and I drove to the remote minimum security camp twenty miles north of Pincher Creek, a collection of tired, tan modular trailers in the lodgepole forest. Brent shot Phyllis nearly within earshot of this place, a place that had become her refuge for the last three winters of her life, and I was ready to know why.

Tommy and I walked up to the trailer with the "Office" sign and knocked on the door. We heard a man say, "It's open," so we went in. Two Indigenous men were sitting on metal chairs. Above the desk was a bulletin board labeled "Current Inmates" with a dozen photos of men. Native Counselling Services of Alberta created this correctional facility in 1980 to offer an alternative to institutional incarceration for Indigenous offenders. Inmates were paid an allowance to work on forestry and maintenance projects contracted through various government agencies. They were also offered traditional Indigenous cultural and recreational activities, sweat lodges, personal counseling, awareness workshops, and a post-release employment referral service. The maximum capacity was twenty men, so it was a relatively small correctional community.

I introduced myself and Tommy to the pair of silent men in the trailer, starting out with: "Hi, I'm Diane Boyd with the Wolf Ecology Project in Montana, and this is my friend Tommy from the University of Montana. We radio-collared and followed a white, female wolf in the Flathead from 1985 through 1988, and then she disappeared from there. I heard that she was hanging around your camp before being shot a couple of months ago. I'm interested in learning why she came here, if she was a problem, how she survived, and if anyone knows anything about her. Can you help me out? We called her Phyllis, and she was all but invisible when she lived in the Flathead."

One of the men said, "She was here, but Gary knows a lot more about her than me." He pointed to his colleague and then headed out the door. Gary hadn't spoken. Into the silence, I earnestly told him a bit about Phyllis's life and legacy as we knew it, then let the conversation pause. Slowly the introductory tension began to unwind, and the story unfolded. Gary and his workmates were members of the Piikani Nation and this was their ancestral home.

They had been hired by the province to provide counseling services for the inmates at the camp.

"Yeah, I saw her a lot over the last three winters. These two big, kinda-wild dogs roamed around a lot here. Sorta mean dogs. Someone in the camp fed them sometimes. The white wolf occasionally ran with the dogs. That worried some people. The dogs bothered campers and hurt the pet dogs at the Castle River Bridge Campground just a few miles up the road. Alberta Fish and Wildlife would get calls about the wild dogs and then they'd pay a visit to our camp and tell us that something needed to be done." During those encounters, he told me, Phyllis was not seen with the dogs. Then the dogs went farther afield and were caught in the act of killing a rancher's steer. Again, Phyllis was not with them.

We all walked outside following Gary as he pointed out the places he used to see Phyllis. "I put out some scraps in bowls outside the kitchen for the wolf. She'd come in at night when she couldn't be seen. I'd see her tracks in the snow, so I knew she had come by. She was hungry." He led us to a big shed that housed the camp's generator, which was rumbling away as we walked through the door; I could feel the heat coming from the engine.

"One really cold morning, the wolf tracks led into this shed and came back out. I looked inside and could see where she had curled up and slept in the warm shed, out of the wind. After that, I propped the door open and left some blankets in there for her. She slept here a lot, curled up on the blankets." He paused. "How old do you think she was then?"

I told him that in 1992, she would have been at least ten years old, which was really old for a wild wolf. "I'm sure your kindness extended her life quite a bit," I said.

I was thinking to myself how this elder wolf, former pack leader, and elusive predator spent the last three winters of her life as a not-so-wild wolf in the shadows of a human camp, aided by a benevolent soul who offered her shelter and food when she fell upon hard times. I imagined that such a connection could be a solace in this harsh, lonely landscape. Clearly, Gary admired Phyllis and looked forward to her visits. Gary said that his grandfather told him stories of hunting on their sacred ancestral hunting grounds all around the camp area. Perhaps, in this place of common origin, something stronger than reason united their lives.

Gary showed us where Phyllis sometimes walked on a path past their sweat lodge. He didn't think that she bothered anybody—just a tired, old wolf looking for a safe place to hole up. We thanked him for his time and insights. Tommy and I drove away, entranced by Gary's unexpected stories and spiritual ties with Phyllis.

THE NEXT MORNING, we learned more about Brent. Dave described him as a successful outfitter for big game and mountain lions, both in the Rocky Mountains and in New Mexico. I felt some contempt for the man who had shot Phyllis, but I wanted to remain open-minded as I filled in this last piece of the puzzle about the white wolf from the Flathead.

A large pickup with several barking mountain-lion-hunting hounds in the back pulled up to Dave's home, and out stepped a tall, wiry, athletic man with rugged good looks. He was about my age, and the crinkles around his eyes revealed years of being outdoors. He greeted us all with handshakes and a friendly "Hi, nice to meet you." Not what I was expecting. He removed his well-worn cowboy hat as we all pulled up chairs around the kitchen table.

"I'm a conservationist sport hunter," Brent said. "Been in the outfitting business for fourteen years. I hunt and guide for deer and elk, but my real passion is lion hunting. I'm not against wolves."

My inner voice overruled my outward calm and I asked, "So why did you shoot Phyllis?"

There, it was out. I had thought about this a hundred times and soon I'd know the real reason. Or at least I thought I would.

Brent took his time laying out the story. He first saw the white wolf in 1988. Her pelt wasn't prime that season. He could have shot her then. He said he didn't because he saw the radio collar and he appreciated the difficulty of catching a wolf to put a collar on it. He was a trapper and had matched wits with lots of cunning wild animals. He had also heard about the research to monitor and recover wolf populations just south of the border. "I think research is important," he said as he looked at me.

But lately, he had heard that the two feral dogs and the white wolf were causing problems with people and pets nearby, and chasing deer and elk on their winter range. "Fish and Wildlife told me the wolf's collar wasn't working anymore. They told me to shoot them all if I had a chance. We were concerned about the wolf and dogs running wild and endangering families vacationing at Castle River. Then, last December, I was out scouting and saw the wolf and the dogs on the road near the camp. The wolf's pelt was in prime condition. I pulled out the rifle and shot her."

"What about the dogs?" I asked.

He left them because a sort-of owner of the dogs was with him when he spotted Phyllis; he didn't want to shoot the dogs in front of him.

"I gave the wolf's skull to Fish and Wildlife for research. I knew she was kind of a famous wolf. The news that I shot her hit the press and it spread like wildfire. I was made to look like the bad guy. I got lots of angry phone calls and it made me mad. I didn't do anything wrong because it was a legal harvest. I knew she was old, so she didn't suffer from dying a slow death of old age. I didn't kill her for no reason." He paused and added, "I'm getting her mounted."

I filled him in on Phyllis's life history as I knew it, from her first litter of pups to her exile from the Flathead, then her wandering around B.C. and Alberta until she settled into her final home near Pincher Creek. We talked for quite a while before he departed with a polite smile and goodbyes all around. I thanked him for his time, and we walked him to the door. Then I fell into a long silence. The two stories, by two different men—one about a special wolf and one about a beautiful pelt—were hard for me to reconcile, especially with my own story.

The facts are that on December 19, 1992, Brent aimed his rifle and killed her with two shots. The first one, in the hindquarters, sent her sprawling out on the snow. Then, a second shot through the chest finished her off. Snow-white wolf on bloodied white snow, and all the wolf wisdom fading from her aged eyes. Her once deadly fangs were reduced to one worn lower canine and one broken upper canine—all the other teeth worn to the gumline or simply gone.

THE LAST TIME I saw Phyllis alive was in March 1989. I had been walking to the outhouse and looking south from the cabin, across the snow-covered hay meadow in Moose City, when I saw a tiny, pale, distant dot traveling north toward me. Coyote or wolf? I quickly turned back, grabbed my binoculars, and determined that it was indeed a wolf, a wolf wearing a black radio collar. I ran to get my volunteer from his cabin, along with our antenna and receiver, and together we slipped between the buildings to hide behind my cabin and watch the wolf approach. We scanned through all the frequencies on our receiver, but none of them pinged in.

I looked more closely as the wolf kept trotting toward us, seemingly unaware of our presence. She was white and her collar was well-worn, unmistakably Phyllis. She continued her deliberate journey across the entire one-mile-long meadow, coming to a stop near the cluster of old homestead buildings less than twenty-five yards from my cabin. She stood there like a statue for a long time, head up, ears forward and erect, staring at my home. Then, she turned her head slightly and spotted me. We locked eyes for a few seconds, before she silently turned around and trotted back in the direction she had come from, exactly retracing her steps. The whole episode lasted no more than ten minutes; then, she disappeared into the willows and was gone, as if she had never been there at all.

Wolves have trotted through Moose City or down the frozen river ice in front of my cabin many times, passing by on their way to somewhere. But why did Phyllis, the queen of concealment, make a deliberate trip straight to my cabin and stop in plain view? Did she recognize my scent, know that this was my house, that she was my ultimate challenge? Or maybe she just dropped by to let me know that she was still out there, still in control, and still smarter than me.

Moose City with old cabins and airstrip in the hay meadow
DIANE K. BOYD

Diane's 1909 homestead cabin at Moose City, midwinter
DIANE K. BOYD

Diane's Toyota pickup, nose-to-nose with a logging truck
DIANE K. BOYD

Diane radio-tracking a collared wolf
BOB REAM

Stony, Diane, and Max MIKE FAIRCHILD

Stony and Max running off wild wolf Sage at Moose City DIANE K. BOYD

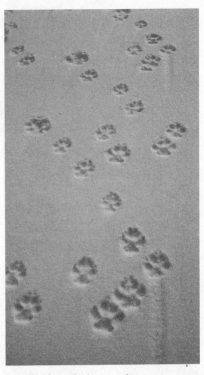

Wolf tracks in snow in the North Fork DIANE K. BOYD

Pilot Dave Hoerner with Diane and her dog Max MIKE FAIRCHILD

Diane and Tom Reynolds dressed for the Polebridge Prom LESLIE WATSON

Paula White, a volunteer on the Wolf Ecology Project, skiing along wolf tracks through dense lodgepole pine
DIANE K. BOYD

Mike Fairchild wading the North Fork Flathead River to follow wolves in Glacier National Park
DIANE K. BOYD

Diane with tranquilized Sage, when she captured the wolf to fit him with a new radio collar PAM BROUSSARD

Diane approaching a six-month-old wolf pup with a jab stick to tranquilize him REGI ALTOP

Diane stopped with colleague to eat lunch in a snowstorm while tracking wolves **DIANE K. BOYD**

Diane—wet, cold, and muddy—setting wolf traps **WENDY COLE**

Diane setting a wolf trap FRANCESCA MARUCCO

Diane with an adult male wolf WENDY COLE

9

...

PEOPLE
AND PLACES

MOOSE CITY WAS more than a collection of rustic cabins, outbuildings, and a rusted 1940s stock truck. It was where I matured professionally and personally, and where I made many memories and lifetime friends.

I had the place almost completely to myself for three years, then, in 1985, Moose City became the headquarters for the revitalized Wolf Ecology Project and I no longer spent so much time alone. At dusk, I walked down to the little cabins to check in with my crew, made sure everyone made it back okay after a day out in the field, and shared in their excitement about what they had learned. Andrea Blakesley and Mike Fairchild would sometimes pull out their guitars and mandolin after work and we'd sing, laugh, and fill the air with music. Andrea and Mike's voices were sweet and harmonious, filling the lantern-lit cabin with sounds of warmth and friendship. One winter evening, in a lull between songs, we heard a wolf howling and we rushed outside—Andrea in her bare feet on the snow. The wolf was answering our music from just across the river. We stood quietly, frosty vapor escaping from our nostrils, as we listened to the rising and falling melodious wolf wails filling the night. The maestro of the singing world had joined in, and it was a mutual serenade.

Dozens of passionate, hardy volunteers and technicians made up the WEP team and were crucial to getting the research done.

Their dedication and can-do attitude were strong and essential. They came from all over the country. Andrea came from Orange County, California, for a quarter of fieldwork—and stayed two years. She loved the work, and we loved her effervescence. We bumped her up from volunteer to technician, which meant she got paid a bit and had some responsibilities over the volunteers. She grew immensely in this role and was the most organized person on the team. After she left Moose City, she moved up to Alaska, where she took a permanent job as an ecologist in Denali National Park, running sled dogs and living a full life from her own log cabin.

Pat Finnegan came from Hawaii and had never been on cross-country skis before he arrived in the North Fork. But he was determined and a quick learner, and after a month of daily skiing and negotiating streams, deadfall, and icy hills, he worked those skis like an old pro.

It was Pat who fished me out of the North Fork Flathead River one warm, late-March day. We waded the high river in the morning when the air temperature was right at freezing. The sun shone gloriously all day long as we skied on wolf tracks out of Sullivan Meadow, wearing just our wool pants and T-shirts. By the time we returned to the river in late afternoon to wade back to the truck, the river had risen just enough to float my feet off the bottom—and start to tip me over and carry me downstream.

Pat grabbed the top of my backpack as my shoulders and long hair were disappearing underwater and pulled me upright by the pack straps. Strong guy. I endured a cold dip that literally took my breath away. When he asked, "Are you okay?" I breathlessly replied, "Ye...ye...ye...ye...yesss!" and we both laughed at my uncontrolled stammering. My chest waders had filled with icy river water, but we finished the wade and skied quickly back to our truck, where I always kept spare dry clothing for Mr. Justin Case. And this was definitely a just-in-case scenario. I was lucky that Pat was with me. Many a skilled fisherman has drowned by falling in a river, where their chest waders fill with water and become a death trap.

The river was a constant source of interesting predicaments, many of which could have—but thankfully didn't—turn deadly. As skiing conditions were difficult one day due to a foot of fresh, heavy powder from an overnight snowstorm, Mike decided to wear his snowshoes instead of skis to follow wolf tracks. But he never reached the wolves' trail. Mike broke through the shelf ice at the river's edge. The fast current grabbed his huge, webbed feet and pulled him down into the turbulent water. Mike scrambled with his hands to get a grip on the ice, clawing for purchase. The water came up to his armpits, and the current tried to pry him off the ice and pull him under the shelf. Mike was alone, but he managed to drag himself out, get back to the truck, and warm up. It was terrifying. He definitely cheated death that day.

It was amazing that in a dozen years of fieldwork nobody broke any bones (except me, that time I was changing a flat), drowned, or had any serious injuries. The WEP crew was scattered over twenty miles in rustic cabins between the Polebridge Ranger Station and Moose City, and we checked in with somebody each evening to make sure we all got home okay. It was a two-hour drive to the hospital, surpassing the so-called golden hour of survival in case of a serious emergency. Everybody carried a radio collar in their backpack so that if they encountered a problem, we could eventually find them—provided they were able to pull the magnet to activate the collar. It was a crude safety net at best. In later years, the National Park Service lent us handheld radios that could usually reach anywhere in the North Fork—or at least reach the park dispatch during working hours. This made our check-ins much more efficient.

Romances happened between volunteers, tempers occasionally flared, but for the most part we got along and worked very well together as a team, helping each other out and making the team stronger. Several of these volunteers and technicians have remained in touch with me to this day. Many have advanced into university professorships or biologist jobs, where they have become leaders in the conservation field. We enjoyed and endured many truly amazing things together, including, of course, our encounters with wolves.

JUNE 20, 1995, was a cold, rainy, miserable day as Wendy Arjo, my PhD colleague; her volunteer, Caitlin; and I checked wolf traps up Tepee Lake Road. The three of us walked a mile behind a locked Forest Service gate with our capture equipment and found a drenched, skinny wolf jumping around in my trap. I walked up to the wolf with my jab stick, and as she tried to get away, I gave her a quick poke in the rump with a syringe to tranquilize her. She went down quickly, and we fitted her with a radio collar and began processing her in the cold downpour. However, her initial temperature was on the low side for an excited, agitated wolf; she was drenched, shivering, and a candidate for hypothermia.

Within five minutes of being tranquilized, her rectal temperature had dropped another degree, so I told Wendy and Caitlin that we must immediately pack this seventy-two-pound wolf out to the truck, start the engine, and run the heater full blast to help her warm up. Temperature, pulse, and respiration rates are critical in monitoring the status of a sedated animal. The trend—going up or going down—was more telling than a single reading. Her temperature was declining rapidly.

Caitlin packed up the tarp and capture equipment, while Wendy and I did our best to carry out this six-foot-long, soggy sack of potatoes in as humane a manner as possible. We made frequent stops to rest and swap positions. By the time we got to the pickup, the wolf's temperature had dropped another degree, and now I was truly worried. The trend was not good. Chilly wolf 9588 had a beautiful black coat and her eyes, instead of being golden yellow—the usual color for wolves—were a striking pale blue, so we named her Ice.

The three of us had completely filled the front seat of the truck when we had driven out to check our traps earlier that morning.

Now we needed to make room for Ice—and there was no back seat. I asked Caitlin to please wait outside the truck in the rain at my driver's side window, so we could all see each other. My plan was to have Wendy and I sit on the front seat with the shivering wolf on our laps so she could absorb heat from our thighs, all while having the heater going full blast. Then when the wolf's temperature had come up closer to the normal of about 100 degrees Fahrenheit and was continuing to rise, I would give Caitlin the thumbs-up. She would open the driver's door from the outside so we could let the wolf out.

We loaded the drugged wolf onto the front seat. Wendy and I crawled underneath her soaked, black body and turned the heater up to the maximum. We managed to finish our processing and get a radio collar on her in cramped quarters. The interior of the truck was now pretty much a stinky steam bath. We had stripped off our wet rain gear and a dense fog was now covering the insides of the windows, but Ice's temperature kept dropping.

"Let's take off our shirts and press our upper bodies across Ice to totally wrap her with our body heat," I said.

So we sat there in our sports bras, hugging this unresponsive wolf sprawled across our laps. We didn't say a word so as not to wake the wolf. I sat in the driver's seat monitoring her eyes and ears, while Wendy, in the passenger's seat at Ice's rump end, took rectal temperatures every few minutes. Poor Caitlin stood dutifully outside in the rain, watching us steaming in the truck.

Wendy indicated with hand signals that Ice's temperature had finally stopped falling; it was now six degrees below normal. We had been warming this wolf in our truck for more than two hours, and finally we had reversed the hypothermia. The wolf's temperature began to rise, she stopped shivering, and my worries began to ebb. Her temperature continued to climb, slowly and steadily. Meanwhile, Wendy and I were about to expire from the heat. I gently reached over the wolf's head on my lap and clicked down the dial on the fan. At the sound of the click, one of the wolf's ears quickly snapped forward toward the sound. Nothing else moved except the telltale ear. I then realized that Ice was awake and aware, even though appearing to be immobile—with her mouth inches from my face.

I vigorously pointed with my head toward Ice for Wendy to see; her eyes grew wide with comprehension. We silently prepared to release Ice, and I gave the thumbs-up to Caitlin watching through my window. She pulled the handle and flung the door open. Ice exploded up from my lap. I grabbed the scruff of her neck with both hands and directed her wildly biting mouth away from me as she lunged for the steering wheel. Wendy braced against her door and vigorously pushed the wolf's rump my way. I couldn't let go of the wolf's neck for fear of being bitten. Wendy pushed harder, I hung onto the wolf, and together Ice and I slid out of the truck and plopped onto the muddy gravel road. Ice jumped up and I let go, and we all watched her wander off, albeit a little drunkenly, into the forest.

Caitlin climbed into the truck, which stank of wet dog and sweaty women. Wendy and I, in our sports bras covered with black hair, looked like we had slept with a gorilla. We watched Ice as she disappeared. She was looking good, and I hope she fared well after her ordeal. When we lost sight of her, Wendy and I got all our rain gear back on, and the three of us then giddily headed out to check the rest of our traps. A month later, I looked out my cabin window and saw a skinny, black wolf in my driveway wearing a radio collar. I quickly grabbed the receiver and antenna from inside the cabin, and I confirmed that it was Ice's signal. Then she looked up at my window briefly. I saw those ice-blue eyes, and I smiled.

OUT IN MOOSE CITY, we rejoiced in the northern lights and meteor showers, with no light pollution to distract from the night sky. Running out to the middle of the grassy meadow, which doubled as our runway, we lay down on our backs on frayed blankets to enjoy the celestial show. The most fantastic northern lights I ever saw occurred on August 6, 1982. I lay on my back on a blanket in the middle of the meadow and watched awestruck as green waves with pink and rose streaks undulated for hours, from the northern horizon upward until they surpassed the zenith overhead. Earlier that day, curmudgeonly, retired university professor Frank Evans had blown his brains out over a failed love, with his former sweetheart and her new boyfriend just a few yards away. God, what an awful and tragic thing that was. Those of us who knew him had no doubt that eccentric Frank was stirring up the cosmos and brilliantly flaming the night sky as his haunted soul left this world.

MY NEAREST year-round neighbor was Tom Reynolds, who had a reputation as an ancient, misanthropic hermit who did not welcome guests. This stirred my imagination; I had to meet him. During my first autumn in the North Fork, John Senger, a young U.S. Customs officer, asked me if I'd like to go with him up to Tom's cabin. Absolutely! We jumped into John's pickup and drove south two miles to Tom's driveway entrance and then another mile up his steep road.

The rutted route wound through mature stands of larch, aspen, and lodgepole and ended at a clearing with Tom's cabin perched high on the ridge, with a stunning view into Canada and Glacier National Park. Smoke was coming out of his chimney, an axe was neatly resting behind the wood chopping block, and his old pickup was parked behind the outhouse. A semblance of tidiness prevailed.

"Tom's nearly deaf, so you gotta speak up real loud," John told me. "We won't go into the cabin unless we are invited in. Not many folks come up here, and Tom ignores most of those who do. He sits in his cabin reading and won't come to the door."

John knocked loudly. I heard a chair scrape on the wood-planked floor as this small, dignified gentleman stood up from his kitchen table, which was piled with magazines, newspapers, and books. Through square glass panes, I could see Tom walking slowly to the white wooden door. I had built up this moment in my mind ever since I had heard Tom Reynolds stories from the North Fork locals. Few had ever met him, but all knew of him. Tom was wearing a black-and-white, buffalo plaid, flannel shirt; a faded silk ascot around his neck; and loose, heavy wool pants cinched up around his trim waist. A smile lit up his face in recognition of John; they were obviously good friends. Then Tom looked at me and his smile remained, for which I was thankful. He opened the glass-paned door and waved us in. He pulled out two well-worn kitchen chairs from his table and waved his hand to indicate for us to sit down, while he grabbed his hearing aids. He lifted them out of their case and pushed them into his ears, fiddling with them amidst tinny squeaks and whistles until he had them adjusted right.

What I learned that first visit, which was followed by countless more, was that Tom was the most interesting person I had met in my entire twenty-five years of life. We grew close in our extraordinary friendship over the years. I considered myself a bit of a loner, but next to Tom I was a social butterfly. Tom chose to live his life at the end of a sixty-mile-long, potholed, nasty road, about as far from other humans as he could get—and that was the way he liked it.

John, hollering loudly and gesturing enthusiastically as he talked, gave his old friend a bottle of Black Velvet whisky. Tom, with his English accent, thanked John, put the whisky in his booze cabinet, and then offered us both a cup of tea. Real English tea, bulk style in a tea ball, steeping in an ancient, delicate, classic white-and-blue porcelain teapot from England.

Tom surprised me in many ways, and I loved his direct honesty, intellect, humor, and wit. After John's introduction, I regularly visited Tom alone, bringing him his mail and my homemade chocolate chip cookies. I learned from Tom that he had been born in England in 1896 to indifferent parents, and his nasty mother had sent him and his brother off to a military boarding school when they were children. The school became his home; he was raised by the school staff and rarely saw his parents. He didn't lament or openly express any emotions about this. He just relayed his background in CliffsNotes style. When he enlisted in the English military, he weighed less than one hundred pounds and stood less than five feet tall. Tom fought for England in World War I and spent time as a prisoner of war. When I asked him about his British upbringing, he snapped, "I'm English, not British!" I didn't make that mistake twice.

After being freed from the war's entanglements, Tom moved to North America, first working in Canada and then landing in the North Fork as the U.S. economy was crashing in the late 1920s. Tom herded sheep up in the mountains and worked for the Forest Service—in the lookouts and fighting forest fires. He bought his 160 acres for back taxes of $109 in 1930, built his unique homestead, and supported himself with woods work in the North Fork. He was a voracious reader of everything, across the political spectrum from *Mother Jones* to *The Spotlight*, as well as *Sports Illustrated*, *Playboy*, *Harper's Magazine*, the *New Yorker*, the *Atlantic*, and dozens of other publications, of which I was eventually the lucky recipient.

THE ANNUAL SHUTDOWN of the Polebridge Northern Lights Saloon was to be a fun formal night in September 1985, complete with young ladies' formals from the 1960s and a disc jockey playing raucous rock 'n' roll. I asked Tom if he would be my date, presuming he would say no, but he smiled and said, "Well, yes," in his clipped English accent. I broke into a big smile and told him it was formal— did he have anything to wear? He got rather indignant and said, "Of course, I have my tuxedo from when I played oboe in the miliary band for England." That was some seventy years earlier, but I guess he knew it still fit. I was so excited to have the honor of escorting this noble and coveted friend to the Polebridge Prom. A month later, I picked him up in my truck for our date, and he stepped out of his cabin in well-fitting black dress pants, black coat with tails, cummerbund, black bow tie, cuff links, and an elegant, ash-gray overcoat. Wow, what a dashing gentleman! It was impossible to believe he was eighty-nine years old.

We arrived at the saloon and a huge crowd gathered to greet Tom; he hadn't been to Polebridge in many years. He smiled, delighted to be in the spotlight. The DJ, wearing a Hawaiian shirt, began playing "Twist and Shout" by the Beatles, and locals gyrated all over the crowded dance floor. Tom shook his head and laughed hard. He said he'd never seen such a thing. No waltzing here. Every woman in the place wanted to dance with Tom, and he smiled and danced all night long, imitating the disjointed steps all around him. His presence and participation was the highlight of the evening for all of us.

ONE THANKSGIVING, I drove up to Tom's cabin to pick him up to join my friends for a hearty turkey dinner. It was snowing hard, and I barely got up Tom's steep, twisty driveway with my pickup in four-wheel drive. I had to take a run at some of the drifts more than once and ram my way through them. When I pulled up to his house, he was dressed warmly, with his snowshoes and kerosene lantern in hand for the return trip, anticipating that his cabin would be snowed in by the time dinner was over. We drove back out, the wipers swiping at the big, heavy, wet flakes. We stomped the snow off our boots as we entered my friends' cabin. The heavy snow fell all through dinner, while my boyfriend and several friends gorged on more good food than I normally see in a week. Tom especially enjoyed the homemade apple pie à la mode, topped with ice cream we had churned in a hand-cranked ice cream maker.

With my belly full and the evening winding down, I shoveled eight inches of new snow off my pickup, warmed up the engine, and then helped Tom get into the passenger seat. We slowly drove the eight miles north to Tom's driveway, the truck mushing around in the deep powder while I peered into the dark past the rapidly slapping wipers into swirling snow. My heart sank because I knew I couldn't drive up Tom's road, and he would need to use the snowshoes and kerosene lantern he had put in my truck. He laced on his old wood-and-leather snowshoes, lit the smelly lantern as he had a hundred times before, and got out.

"Tom, can I walk with you up to your house?"

"No, I'm fine, thank you."

And then he headed off into the darkness, illuminating the big, white flakes with his lantern. I watched the diminishing, swaying light disappear into the falling snow as the old man made his way home alone, as he had always done.

I skied up to his house the next afternoon to make sure he had gotten home all right. He had. He told me that he snowshoed down to greet Becky, the mail lady, at his mailbox that morning. En route, he tripped on his snowshoes and fell in the deep snow. As he wallowed around with his bamboo ski pole, struggling to get to his feet, he looked up and saw a large gray wolf watching him. He yelled at the wolf, telling it to go away. The wolf stared at him and then disappeared into the forest.

"If he had been hungry, I wouldn't be here to tell you the story," relayed Tom matter-of-factly.

"If he had wanted to kill you, Tom, he sure had his chance, hungry or not."

Tom thought about that for a few seconds and stashed the thought away. He had no love for wolves or other predators. Then we shared a cup of tea and fresh bread baked in his wood-fired kitchen stove.

Seven years later, pneumonia took Tom at age ninety-six in December 1992. He was still hauling water in buckets, splitting wood, and washing his laundry by hand right up until the week he died. The mail lady, Becky, was concerned because he hadn't come down to greet her at his mailbox like he always did. When she arrived at his cabin, there was no smoke coming out of his chimney. She went inside and it was very cold. Tom was lying in his bed at the back of his cabin, unaware that she had come inside. Becky saw him pick up his alarm clock and look at the time. He must have known that Becky, not finding him at the mailbox, would be there soon, for he placed the alarm clock back on the night table and relaxed, waiting. Becky rushed to him and told him she would go get help, but he signaled her to wait with him. He tried to speak but couldn't. Tom drew his last breath in his cabin, in his bed, and chose to die with a dear friend sitting next to him, rather than alone. Rest in peace, my special friend. You lived and died well. I miss you still.

BY THE LATE 1980S, the WEP was following four packs, and Mike Fairchild and I were kept busy trying to keep up with the growing wolf population, with the help of a half dozen adventurous volunteers. In winter, my favorite season, we spent every day radio-tracking wolves. We triangulated their locations on paper maps using their signal directions and compass readings and then skied in and followed them, backtracking on their trails and collecting data on how they were making their living.

Glacier National Park allowed us to stay in their backcountry cabins for our work as long as the cabins weren't already being used by park staff. My favorite wolf place was Kintla Lake and the little ranger cabin there, built in the 1930s. Getting to the cabin in winter involved skiing to the North Fork Flathead River; wading across it with a backpack stuffed with food, sleeping bag, and skis; and then skiing five miles up to the lake. The cabin was another one-third mile on skis from the southeastern lakeshore.

The one-room cabin was kept ready for park staff and visitors like us. Wood was stacked along the outside wall, along with a full box of kindling. The picture window looked out over the stunning lake for most of its five-mile length. The Murphy-bed-style table was the exterior wall of a cabinet. When you pulled down the cabinet wall, a leg swung out to support the tabletop. It was a bit rickety but perfectly placed beside the picture window. Sometimes we saw deer or wolves out on the frozen lake while we ate our breakfast—which was somewhat precariously balanced on the one-legged table. And sometimes when I went outside to pee at night, I saw the northern lights dancing over Starvation Ridge. At daybreak, I slipped on the heavy leather ski boots I'd left warming by the woodstove overnight and ventured outside to knock the ice out of my ski bindings. Then, I stepped onto my skis, clicked into my three-pin bindings—as did my colleague—and began swooshing out onto the skiff of snow covering the frozen lake. We always stayed in backcountry cabins with another person for safety—and two people could cover twice as much mileage as one person could along different wolf trails.

Wolf and mountain lion researchers were often based out of Kintla Lake in winter because it was a wildlife haven. The microclimate reduced snowfall, which made the lake valley a coveted winter range for deer and elk, with a corresponding abundance of their predators. The lake itself was usually covered with a thick lid of ice with a dusting of snow on top, which made it perfect for following deer, wolf, and lion tracks—and heavenly for skiing.

Sometimes at first light, we saw wolves out on the ice feeding on a deer that they had killed on the slippery surface overnight. We had to wait in the cabin, with binoculars and hot cocoa, watching the feeding frenzy out the door. After the wolves left, we could ski to the kill, collect the jaw for aging, and do our csi forensics.

We collected incisor teeth from adult ungulate jaws and sent them to a lab for cementum annuli processing to accurately determine the age of the animal. Ungulates generally add one layer of growth to their teeth annually, and in cross section the teeth show rings, like trees; you can count the rings and accurately determine the animal's age. The overall wear pattern in adult deer will put you in the ballpark in determining age, but this estimate isn't as exact as using the cementum annuli aging technique. Fawns and yearlings are easy to classify in the field because of the tooth eruption pattern, which is when permanent teeth replace deciduous (baby) teeth. We also assessed the dead animal's overall condition based on the marrow fat content.

One morning, when I was way out in the middle of the lake, I heard whale songs and stopped. Was it distant howling? It came again, accompanied by eerie creaks, groans, and sonar submarine pings. The ice was acting like a membrane, reverberating the sounds up and down the valley. Then I realized the elemental music was coming from the ice itself as it contracted and expanded, very much alive. I knew the ice was two or three feet thick even far from shore. You could land a plane on it, if that wasn't illegal. But hanging out on a sheet of ice suspended over four hundred feet of water was somehow scarier than being on noisy ice over ten feet of water. After my moment of panic, when I realized that I was not about to fall through the ice, I relaxed and took it all in. Then the wolf tracks beckoned, and I was soon off on another day of adventure in one of the most magical places in the park.

MOOSE CITY TOO was a magical place that I fell in love with from the very start. But as my years in the North Fork stretched from a master's thesis to several more fulfilling years with the WEP, I began to long for a home of my own, on my own land, with my privacy. I looked for several months until I found a beautiful and remote parcel of land with a mountain view, a spring-fed creek that flowed year-round, moose-browsed willows, and grizzly bear scats along the game trails. My new neighbors were four-legged—or if two-legged, they also had wings. I quickly fell in love all over again. I made my first payment on the property in 1982 and envisioned building my cabin on this densely timbered land, without even a driveway cut into it yet.

I bought a small, historic homestead cabin that had been built in 1910 and, with the help of friends, took it down log by log in four days, moving the logs by hand onto a flatbed trailer. I numbered each log with a piece of duct tape that I attached with a heavy-duty staple gun. Additionally, I measured each log and took photos of every standing wall before tearing down the ancient cabin. Then we hauled the logs twelve miles to my homesite and unloaded them in a pile, where they sat under a black plastic tarp for two years. It took me that much time to save enough money to have local master builder Ron Wilhelm build a proper foundation with concrete blocks and a footing sunk three feet into the ground to reach below frost line. Some of my log tags had come off under the plastic tarp, but with the help of my notes on the original measurements and photos, I could easily identify each log and retag them correctly.

I held a huge log raising party, with enough burgers and beer to feed the large group of friends who showed up to help raise the walls in one long day. Over the next seven years, with the help of many more wonderful friends, I slowly rebuilt my own dream homestead cabin out of the original old logs, and added a full upstairs and a breakfast nook with a skylight that the morning sun streams through. It took me that full seven years of rare days off from the WEP, and scrimping from my small WEP salary, to get the cabin completed. In September of 1991, I moved out of Moose City and into my own cabin, at last. I continued to work for the WEP and to commute from my homestead for several more years. Tucked into my home, I fell asleep to the rush of the creek below my bedroom window. I had accomplished a huge undertaking, always sure I could do it even though I didn't have a clue how to build a cabin, plumb propane lines, install the metal roof, or set the windows. I was young, strong, idealistic, and so very gratified.

10

......

FROM BASS CREEK
TO SPOTTED BEAR

W HEN I STARTED working for the Wolf Ecology Project
in 1979, I was working on my master's degree, which I
completed in 1982. I continued working for the WEP for
another dozen years and then worked on my PhD from 1993 to 1997.
Analyzing and writing up fourteen years of data was a huge task—
no doubt the longest running set of field data the University of
Montana had ever encountered for a PhD project. Funding ceased
for the WEP in 1995, so graduating as "Dr. Boyd" helped prepare me
for the next stage of my career.

As wolf numbers and conflicts continued to increase, I realized that in addition to doing ecological fieldwork, the key to wolf
recovery success was working with people. Finding ways to prevent or resolve wolf-livestock-human conflicts and trying to change
misperceptions of hunters became a more important part of my
post-PhD work. In the late 1990s, I worked briefly for the U.S. Fish
and Wildlife Service under Ed Bangs. I was based in Helena, and
my job was to monitor wolf recovery in northwestern Montana
and work with ranchers to help reduce conflicts with wolves.

By the late 1990s, the wolf population in the North Fork had
overflowed, and dispersing wolves had set up territories far from
Glacier National Park. Wolves were being seen, heard, and pho-
tographed in new places in western Montana. It was great news

for wolf recovery efforts, but not such great news if you were a rancher and wolves were seen among your calves and sheep. When these early dispersers started to tangle with livestock, the idea was to catch and translocate them to more suitable territory because, despite the successes, there were still so few wolves in the 1990s that each one was critical to the recovery of this endangered species.

A solitary male wolf crossed paths with a single female wolf as they roamed the Bass Creek drainage of the Bitterroot Valley, Montana, in the winter of 1998–99. There was plenty of time for them to get acquainted before the peak of breeding season. Soon, their double scent-mark urinations and accompanying long-legged scratches sprinkled the landscape, advertising visually and aromatically that this was their new territory. Much flirting, nuzzling, tail wagging, playing, and curling up to sleep together helped strengthen their bond—not so different from humans. Howling and tracks additionally served to warn potential trespassing wolves to keep out. But it was unnecessary, since wolf howls had not echoed through the Bitterroot in seventy-five years.

WHEN THE BASS CREEK wolves found each other, the population of wild wolves in Montana comprised ten breeding packs and a few loners, adding up to approximately eighty endangered wolves in the state. The Bass Creek pair was a tenuous strand of the southward-expanding network of wolves. Twenty years had passed since the arrival of the first recolonizing wolf, Kishinena, near Glacier National Park, and the growth rate of the Montana wolf population had plodded along in fits and starts ever since.

In the 1980s and early 1990s, as each new wolf pair coupled up and had pups, tragedies befell the emerging packs. There were some natural mortalities—like being killed by prey, competing predators, other wolves, starvation, and avalanches—but most deaths were caused by humans. Parvovirus and distemper, likely transmitted from people's dogs, killed entire litters of pups. Of the four packs that I had radio-collared in Glacier National Park and British Columbia, two lost all their pups in the spring of 1989 to parvovirus. But more often, the cause of death was poison, vehicle strike, illegal poaching, or predator control to alleviate losses of cattle and sheep.

The low density of wolves in Montana made it slightly more challenging for a lone wolf to find a mate. However, wolves can easily travel thirty or more miles in a day, and mate-seeking wolves will eventually cross paths—if they aren't killed first. They are born distance runners with lean, narrow-keeled bodies, long legs, and a ground-eating stride that is a model of efficiency. A wolf is fed by its feet—so goes an old Russian proverb. Five-inch-long paws float these canine traveling machines across deep snow where prey may flounder. Crossing mountains and rivers doesn't faze them, even in the dead of winter.

THE BASS CREEK pair bothered no one at first and was only detected because the 108-pound male was accidentally caught in a coyote trap on Tom Ruffatto's cattle ranch near Stevensville on December 30, 1998. The fur trapper, Fred, immediately recognized that he had an endangered wolf jumping around in his trap and called Montana Fish, Wildlife & Parks to report it and get some help to free the wolf. Carter Niemeyer, a federal Animal Damage Control wolf trapper from Idaho, happened to be in the state game warden's office in Missoula when Fred called. As coincidences go, Carter was there to search for wolf tracks in the Bass Creek area preemptively, in case there were any wolf-livestock problems.

Carter and the game warden headed to where Fred, the trapper; Tom, the ranch owner; and the wolf were waiting for them. Carter noosed the wolf with a catchpole and hand-injected a tranquilizer into its rump. He put the wolf in a culvert bear trap to hold it for radio-collaring. Carter called my colleague Joe Fontaine at the U.S. Fish and Wildlife Service office in Helena and asked him to bring a radio collar. Joe came over the next day with a collar, and they released the wolf on the neighboring Ed Cummings ranch the next day.

After the wolf ran off, Ed invited Carter and Joe for dinner at his ranch that night. Around the dinner table, Ed came up with the idea to build a device that could pick up the radio-collar signal when a wolf was near a calving area and then make some kind of racket to drive the wolves away. It might just keep the wolves and cows safely separate. The U.S. Fish and Wildlife Service ran with that idea and the enthusiastic, brainy John Shivik, a researcher with Animal Damage Control, developed the radio-activated guard box, aka the RAG box. The RAG box succeeded in accomplishing what Ed had dreamed up: picking up the radio-collar signal when the male wolf was within a quarter mile of the RAG box and the cows were in the calving pasture. The RAG box then exploded into life, sending out a barrage of flashing lights and a wailing siren, which scared the wolves away. This device was later modified in other locations with different soundtracks, including the loud whop-whop-whop of a helicopter, gunfire, screeching of car brakes, or country western music, with the latter being particularly effective. Shivik worked his geek magic and helped save a few wolves in the process.

NOBODY KNEW THEN that there was a female wolf in the territory or anticipated the saga that was about to unfold. The Bass Creek pair denned less than a mile from the Ruffatto ranch and produced eight healthy and hungry pups in April. The Ruffattos were a long-time Montana ranch family used to figuring out how to handle challenges, but wolves were definitely a new wrinkle in the family cattle operation. At first, the Ruffattos were tolerant of the novel wolves and worked with the U.S. Fish and Wildlife Service to head off conflicts. The Service gave Ruffatto a telemetry receiver so he could monitor proximity of the radio-collared male to their cattle. It was strategically wise to involve the landowners in monitoring the situation, plus it gave the Service an opportunity to meet with them regularly.

Some of the neighboring ranchers took daily horseback rides up into the woods in the general vicinity of the den to help keep track of the wolves and the cows. None of these ranchers had probably seen a wolf before, and certainly not a pack of them in their back-yards. The ranchers weren't hostile, but they had a vested interest and told the Service what they saw. The wolves responded to the RAG box as intended, and they didn't bother the calves for a while. But eventually, the rancher had to move the cows and calves from the confined calving pasture to the open range, where it was impos-sible to effectively use the RAG box among the widely dispersed herd. It was then that the Bass Creek Pack began killing calves.

JOE AND TED NORTH (another Animal Damage Control trapper) had pinpointed the den location during the concurrent wolf pup-ping and cow calving seasons. In early June, Ted re-trapped the adult male, and the wolf pulled the trap away through the brush and out of sight. Ted and his dog, Chief (a Jagdterrier, which is like a Jack Russell terrier but tougher), combed the undergrowth searching for the trapped wolf. Chief found the wolf sitting up to his shoulders in water in an irrigation ditch. He was soaking wet and still caught in the trap. Chief roared up to the cornered wolf for the takedown. The wolf grabbed that feisty, twenty-pound dog and flung him through the air; Ted was sure his best buddy was dead. But that fearless little terrier instantly ran back for more. Ted had to control Chief with a stick to keep him away from the wolf. The difficult part now over, Ted tranquilized the male wolf and removed him from the area to hold him in captivity nearby until they could catch the rest of the family.

The adult female wolf was cagey and catching her proved to be more challenging. Ted was an excellent trapper and knew how to place a trap so that in hundreds of square miles of wolf territory, a wolf would put its foot exactly on the trap trigger. Ted finally captured the lactating and underweight female in a blind trail set, which is a trap set in the middle of a trail without any lures, the ultimate in camouflage. She was already wearing a radio collar. Records showed that she had been captured as a yearling in the Murphy Lake Pack in 1997 up in northwestern Montana, and the collar ceased transmitting there in March 1998. She had traveled at least 150 miles from her natal pack to become the breeding female of the new Bass Creek Pack. Ted fitted her with a new radio collar and released her on-site near Bass Creek to return to her pups, hopefully increasing the odds of eventually capturing the entire family group.

Unfortunately, after the male was trapped, the female shifted her interest from calves to adult cattle. The pressure was on to quickly capture all eight pups and their evasive mother. She now moved her pups a mile up a mountainside to a more remote rendezvous site. Kent Laudon and Isaac Babcock, Nez Perce tribal biologists from the wolf team in Idaho, joined the tracking efforts to help keep tabs on the wolf family. By this time, there were about equal numbers of wolves and humans on-site, with the humans trying to figure out how to capture the wolves and manage this situation.

JOE, TED, AND CHIEF went to the den to capture the two-month-old pups. The mother wolf was gone when the would-be rescuers approached the den, and Chief ran himself ragged trying to bay up the pups, who were scattered around in several smaller holes. Chief worked furiously in the heat, panting heavily, and had to rest. Then the pups ran down into the main den hole and to the very back, creating a new challenge to get them out. After a while, Chief cooled off and charged down the den hole, where he encountered a growling mass of wolf pups about his size. Rather than grabbing them one by one and pulling them out, Chief was holding them at bay in a standoff.

Joe was at the den entrance right behind Chief and hollered back to Ted, "Chief is blocking the entry and he isn't coming out."

"Grab Chief by the tail, pull him out, and toss him away from the den."

All right then. Joe grabbed Chief's stub tail, yanked, and pulled the stubborn dog backward and tossed him toward Ted. But before Joe could dive down in the den hole, Chief raced back, shot between Joe's legs, and was again blocking the entrance. Joe repeated the Chief yank-and-toss several times. Finally, Joe pitched Chief through the air far enough away that Joe won the race. Joe groped around in the blackness of the den and pulled out three pups, one at a time, by the scruff of their necks. Once in hand, they were gentle, docile, and unabashedly cute.

Over the course of two weeks, all eight pups were safely caught and reunited with their father in captivity in Idaho. Toward the end of the pup captures, Ted trapped the mother wolf far from a road, and the neighboring rancher offered to carry the sedated female out on horseback, draped over his lap. The horse wasn't thrilled about this arrangement, but it packed out the sedated wolf. Mama wolf was held temporarily in a wolf crate in a nearby barn and was soon reunited with her eight pups and her mate in their new Idaho pen. The success was a team effort, but the capture of the Bass Creek Pack was due in no small part to the compact but mightiest player on the federal team, Chief.

THE MOTHER, FATHER, and their pups were moved together to a remote holding facility near McCall, Idaho. Here the family group would be safe and well fed, and the pups could grow up without encountering livestock. The goal was to move the family back to the wilds of Montana in December when the pups were larger and would have a better chance of survival if the adults disappeared.

Unfortunately, several tragedies occurred in captivity. The adult male was being treated for a foot injury he had sustained during his

Bass Creek capture and was accidentally strangled by a catchpole inside the enclosure during a capture attempt. After this fatality, the handlers always carried a pair of cable cutters in their back pockets to prevent a repeat. Three of the eight pups then contracted parvovirus and died in August. By the time their release trip was to begin, the Bass Creek Pack was reduced to the mother wolf and five surviving pups, two males and three females. Could a lone adult female feed five growing pups and provide the social structure needed to keep them together? Nobody knew, but wolves are resilient, and all hoped for the best.

The U.S. Fish and Wildlife Service staff built a temporary holding pen made of a soft-mesh electrified fence at the Spotted Bear Ranger Station, on the edge of the Bob Marshall Wilderness in northwestern Montana—an area where there was no livestock and no established wolf packs. Maybe the chance of success was a crapshoot, but it was definitely a more strategic release than the majority of failed earlier translocation attempts, where wolves were literally shaken out of crates before blazing off toward the sunset in a totally unfamiliar landscape. In those cases, only 20 to 25 percent survived more than a few weeks, and almost none lived long enough to reproduce. Most translocated wolves died due to humans causes, with government wolf control being the greatest source of mortality. Surely, we could improve on that tragic record.

On December 6, Joe picked up the six crated Bass Creek wolves in Idaho and drove them two hundred miles east to meet me in Lolo, Montana. We slid the crates from his truck to mine at 9 PM. I was too tired to continue the drive to Kalispell for their transfer to their temporary wilderness holding pen, so I called my dear friend Helen Bolle in Missoula (half an hour away) to ask if I could spend the night at her house, apologizing for the lateness of the hour. At eighty-four years young, Helen was a real fireball and a wolf lover. She had followed the saga of the Bass Creek wolves over the course of the summer and lamented their fate.

"You're always welcome here, dearie. C'mon up, I'll have the lights on."

"Helen, I've got six friends with me."

With hardly a pause, Helen replied, "They're welcome too. I have lots of blankets and couches."

"My friends all have four feet." Helen paused and asked me who I was bringing.

"The Bass Creek Wolf Pack."

"Oooh, hurry up, dearie!"

I headed north on Highway 93 to Helen's, pulled into her driveway along Greenough Park, and shut the engine off. It was cold and snowy, but she came out in her big snow boots, with her winter coat pulled over her nighty and long johns. She peeked into the back of my truck, in awe of the wolves cowering in the back of their crates.

"Oh, they're so beautiful; I feel so sorry for them. Do you think they will be okay?"

"Yes. They'll sleep off the trauma of the day in the cool darkness. All will be well, unless they start howling and wake up your neighbors."

"Oh, I hope they do!"

I put small ice chunks in their crates for water, and then Helen and I headed into her warm house and enjoyed a cup of hot chamomile tea. Soon we fell into exhausted sleep, as did the wolves. No howling was heard except in Helen's dreams.

THE NEXT MORNING, I drove my canine cargo the two hours to Kalispell to meet Joe and Dave Hoerner, our trusty bush pilot during our WEP years. Joe and I were all set to tranquilize the wolves and load them into Dave's four-seater plane-on-skis for backcountry winter landings. Dave had removed the plane's seats to accommodate the six gangly, drugged wolves. The tranquilized wolf family was laid out like sardines, backs to bellies, completely filling the seatless cabin of the small plane.

Dave and Joe flew the wolves into the remote, snowy airstrip at Spotted Bear Ranger Station, with extra tranquilizing drugs at the ready just in case. Tom Meier, a colleague at the U.S. Fish and Wildlife Service office in Kalispell, and Kent, from the Idaho wolf team, had driven into Spotted Bear the day before and packed the primitive runway with a snowmobile so the ski plane could land. While Dave and Joe were flying the six drugged wolves the half hour into Spotted Bear airstrip, I drove eighty rutted, icy miles into the Spotted Bear Ranger Station to meet up with everybody and stay overnight with the wolves.

Tom and Kent met the planeload of wolves on the snow-packed Spotted Bear airstrip.

"There's big old tracks of a single wolf trotting down the edge of the runway," Dave exclaimed with surprise as he climbed out of the Cessna.

"Wow, who could that be?"

We didn't know of any wolves in this wilderness location. Tom, Kent, and Joe loaded the limp wolves into the back of an open pickup, wolf urine sloshing around under the wolves, and drove the three miles from the runway to the pen, with a lot of wolfy scent dripping out of the tailgate along the truck tracks. They carried the six groggy wolves into the pen, then went back to the cabin and warmed up by the crackling woodstove while Dave and Joe flew back to Kalispell.

The wolves had all been loaded into the holding pen by the time I arrived, and they had begun exploring their newest home. With binoculars, we could see the pen from the cabin, but we weren't so close as to be a disturbance. This pen would be their new home for a few days of acclimation to their new environment while we babysat them from the cabin. After dark that evening, when the wolves had awakened and were pacing nervously in their pen, we heard howling—lots of howling—that continued throughout the night. Higher-pitched pup howls mixed harmoniously with deeper adult howls that echoed off the canyon walls and rebounded off tree trunks—distorted and moving, sounding like a dozen wolves. We would soon learn that there had indeed been one wolf on the prowl right outside our door that night.

At daybreak the next morning, Tom was up, dressed in wool pants and a heavy flannel shirt, and making cowboy coffee on top of the woodstove, smoky puffs merging with the pleasant aroma of a morning brew. I opened the cabin door and quietly stepped outside into the soft snow with my binoculars in hand. The peaceful white landscape was silent except for the whine of a nervous pup. I shuffled about thirty feet in my winter boots, when I came upon immense, fresh wolf tracks outside of the pen. Uh-oh.

I scanned the pen with my binoculars, worried that one or more of our wolves had gotten out. One, two, three, four, five, six. They were all inside the pen, and I didn't see any damage to the mesh walls, so nobody had escaped. I raced back to the cabin breathlessly and said, "We've got a seventh wolf out there!" We figured out that the big-footed wolf on the runway, certainly a male, had followed the musky, urine-scented trail to the pen. It was the perfect Christmas party invitation from a ready-made pack missing an adult male.

I aimed the antenna and receiver out toward the pen, scanning for the frequencies of any radio-collared wolves that could possibly be here. The static crackled as I flipped from one frequency to another from our master list of all radio-collared wolves in northwestern Montana. Suddenly, a signal loudly pinged out from very close range. Bingo, the mystery male was 117. This big boy was last located forty miles away a month earlier in the Seeley Lake area. He had killed cattle and had been translocated into this area eleven months ago from Pleasant Valley, near Marion, eighty miles to the west.

We called Joe in Helena, had a quick discussion, and Tom and I quickly agreed to cut short the wolves' acclimation pen stay and turn them loose, pronto. We rapidly approached the pen and opened the door; we also cut a large hole in the mesh with knives to give the wolves several escape options and then backed off to watch with binoculars. The mother wolf left immediately, followed forty-five minutes later by one pup. The other pups were confused and hesitant to leave, so Tom and I hazed the remaining four pups until they ran through the opening and joined their freed family. The male was apparently waiting just out of sight in the forest, as we heard his deep booming howl amidst a cacophony of excited wolf howls. These seven wolves had just become the Spotted Bear Pack. Sometimes Santa drives a sled, and sometimes he flies a Cessna.

THE WOLVES HUNG AROUND within earshot of the cabin, howling all day until well after midnight. Sometime during the night, as we slept in the toasty cabin, all seven wolves headed together up the South Fork of the Flathead River into the Bob Marshall Wilderness. I pictured the pups testing and submitting to their new male leader, while the adult female and male, tails high, began the spirited ritual of becoming mates. They stayed together as a pack, successfully killing wild game all winter, and defining their new territory. The new pair produced a litter of pups that spring, and their now-yearling pups stayed with them—but we had no idea if the pack would stay put or survive.

Previous wolf translocations in Montana and Minnesota had resulted in many terrible injuries and deaths, and they were largely unsuccessful in reducing domestic livestock losses. Wolves had a tendency to return home or move to a habitat similar to the one they had come from—countryside with cows and sheep. Translocations were the first attempt to resolve the problem, but repeat offenders were sometimes killed outright. The issue was complicated and new to the agencies involved. This was the first naturally recolonizing wolf population in the western U.S. The Endangered Species Act required agency management to help wolf populations recover, but the powerful livestock industry did not want wolves—period.

Those in charge of managing wolf populations believed that if livestock killings were quickly addressed with wolf removal, greater tolerance would be created for the predators on the landscape in the long run—kill a few offending wolves and leave non-depredating wolves alone. But there was no way to scientifically test this by leaving livestock-depredating wolves on the landscape and seeing what happened—it wouldn't be tolerated by the ranchers. The pro–wolf recovery public said this was hogwash. It was critical that the wolves survived long enough to reproduce and boost the small wolf population. The U.S. Fish and Wildlife Service admitted in their 1999 annual report: "Wolf control in response to livestock depredation may be a factor limiting the expansion of the NWMT [northwestern Montana] wolf population."

FAST-FORWARD TO MAY 2016—my first year as the wolf and carnivore specialist for Montana Fish, Wildlife & Parks. My job took me throughout the northwest corner of Montana to capture, radio-collar, and monitor wolves, and I worked with hunters and the public to give them good wolf information. In mid-August, I coordinated with Forest Service staff working at the Spotted Bear Ranger Station to bunk in their cabin and try to capture and collar a wolf from the Spotted Bear Pack.

Feelings of déjà vu crept into my mind—the only other time I had been here was that very cold December in 1999 when we had released the former Bass Creek Pack into the wilderness, which had then morphed into the successful Spotted Bear Pack. Three days after I set out my trapline, I captured two males: a yearling and a robust adult with a lot of attitude who was undoubtedly the breeding male. The adult wolf was enormous and confident, had large testicles, and was probably a descendant of that original Bass Creek female—or some second cousin-in-law three times removed.

I was thrilled to see that the pack's legacy had survived from their shaky beginning. The Spotted Bear Pack never got into trouble and remains there to this day; we finally had a successful translocation of former livestock killers who had thrived after human intervention. The story of the Bass Creek to Spotted Bear translocation was extraordinary because of its rare, lasting success in early wolf recovery.

By 2010, wolves in Montana were no longer on that precarious leading edge of a recovering endangered species. Instead, hundreds of wolves now left tracks and scats in every drainage in western Montana. Now that they were no longer endangered, they could be legally shot and trapped in Montana and Idaho. Deep challenges remained for maintaining wolves on human-dominated landscapes among a strongly divided public. But in Montana, wolves had finally hit critical mass and repopulated the landscape under their own productivity and prodigious power, sometimes despite our human efforts to manage them.

11

·····

MY EUROPEAN
VACATIONS

HILE WOLVES WERE slowly spreading through western
Montana in the 1980s, a similar phenomenon was happening with wolf populations in the mountains of Italy,
Poland, Switzerland, and Romania. Remnant populations of the
Iberian wolf in northwestern Spain and northern Portugal grew.
Wolves walked from Finland into Sweden in 1978 and from there
continued west to Norway. Germany, the Netherlands, and even
Denmark would become home for dispersing wolves. There were
no reintroductions in any part of Europe—the wolves were able to
recolonize their former habitat through dispersal, enabled by legal
protection from humans. Humans occupy even the wildest places in
western Europe. Sheep graze in the national parks; shepherds live
nearby and herd livestock up onto the very top of the mountains;
forests are heavily managed for products and human use. And yet
wolf populations have managed to squeeze into human-dominated
landscapes and survive.

Like wolves in the U.S., wolves in Europe were heavily perse-
cuted and extirpated through most of their former ranges in the
1800s and early 1900s. In both North America and Europe, public
attitudes shifted from persecution to protection in the 1980s and
1990s. This set the stage for the wolf's return to its former range,
and conservationists as well as livestock producers soon wanted to

know what was going on with wolves. I connected with German biologist Christoph Promberger and Romanian biologist Ovidiu Ionescu at a professional international wolf conference. As I had worked with returning populations in Montana, I was privileged to be hired to help these two biologists on their early wolf research programs in Romania.

I ARRIVED AT the Munich airport in the summer of 1994. I had been hired to teach Christoph and Ovidiu how to trap and radio-collar wolves in the Transylvania region of the Carpathian Mountains, the wildest and most beautiful mountains in Romania. Christoph was a biologist of endless optimism, enthusiasm, and curiosity. He had secured funding from the Munich Wildlife Society to support the Carpathian Large Carnivore Project and get some wolves radio-collared. Ovidiu was a friendly Romanian but more reserved—although just as determined to see this project succeed.

I immediately liked both of these dedicated men, who perse-vered through challenges in that country that complicated even the simplest tasks. The now-familiar term "supply chain shortages" was an incessant and accepted way of life there. Our Suzuki four-wheel-drive vehicle stopped running due to a clogged fuel filter after we had unknowingly filled the gas tank at a gas station with dirty gasoline. We couldn't buy the $4 fuel filter that every auto part store in America carries. Instead, we put the Suzuki up on blocks, drained the fuel tank, filtered the gas through T-shirts, vig-orously blew out the fuel filter ourselves several times, siphoned the cleaned gas back into the gas tank, put everything back together, and hoped it ran. It did!

Nicolae Ceaușescu, Romanian dictator from 1967 to 1989, was a brutally oppressive tyrant. Under his rule, his citizens suffered great deprivation while he and his wife, Elena, lived lavishly. Nonetheless, the world was shocked when on Christmas Day in 1989, Ceaușescu and his wife were tried by a drumhead court-martial that lasted one hour. They were dragged into a small courtyard on a military base, lined up outside against a toilet wall, and shot by a firing squad.

When I visited, Romania was still grappling with over two decades of severe oppression and economic collapse. I was unaware of the bloody struggles and transition of power that preceded me by five years, but I would feel its lingering effects during my visit. No wonder the people were untrusting and fearful of our team—a German man, an American woman, a Romanian government man, and Christoph's blue-eyed husky—as we drove around the countryside in a jeep-like Suzuki.

I quickly learned that in Romania the only time a woman drove a vehicle was when her husband was too drunk to drive. Whenever I went out walking around our little village, Prejmer, everybody stared at me. If I waved and said "bună ziua" ("hello" in Romanian), they moved away. They apparently hadn't seen many tall, American, blond women dressed in blue jeans walking alone. Prejmer had no tourists to experience the picturesque, old, stony streets and buildings, the horse-drawn wagons, and the rustic lifestyle. I felt like I had gone back a century in time. The dark-haired and short-statured village women mostly wore skirts, and they were always with a knot of children or other adults. Most people in this rural community did not own a car, and if they did it was an old, beat-up, Romanian-made Dacia car—not an expensive, foreign, four-wheel-drive Suzuki.

Ovidiu told me that in 1994, the average Romanian salary was roughly equivalent to US$100 per month, and farmers probably earned less in our small village of Prejmer. The economy was unstable, with the inflation rate between 1991 and 1995 averaging 161 percent per year. Most rural Romanians in this region moved by horse and wagon or bicycle. The more prosperous farmers owned ancient tractors. Hay was harvested with a scythe—a curved wooden blade with a handle—with a farmer swinging the long blade through the standing grass. Family members would come along with pitchforks to move and stack the hay, and eventually load it up onto horse-drawn carts.

Our home was an upstairs apartment owned by the old woman below us. It looked out onto her yard of chickens and one milk cow. The gnarly old rooster chased me one morning, and I had to defend myself with a long stick. I wanted to cook that bird. Christoph had modernized and furnished the apartment to German standards, so we were quite comfortable. One of the wooden planks was broken on the outside steps leading to our abode. Every time we walked up or down the steps, we had to avoid falling through the hole, especially after dark. Christoph asked the old woman, in Romanian, if he could replace the broken plank and fix the step. She stared at him in disbelief and replied, "You know where the hole is—don't step there." She had lived with so little for so long that she couldn't fathom wasting precious money on such a frivolous repair. Christoph quietly repaired it and hoped to not offend her.

We set traps along likely wolf travel routes in the mountains, with me showing Christoph and Ovidiu about trap placement, lures, and set camouflage. Peasants, shepherds, dogs, sheep, cows, pigs, wolves, roe deer, and brown bears all used the same travel routes. With all this traffic in a relatively restricted area, the challenge was to try to catch only wolves. We often ran into shepherds and their guard dogs moving sheep around the wild mountain pastures. The smaller herding dogs were not intimidating, but the large guard dogs wore huge spiked collars to deter wolf attacks and they were viciously protective. Ovidiu told us to always walk with a long staff to fend off the big dogs. I love dogs, but after our first encounter with one of these half-wild guard dogs, I understood and complied.

One shepherd told us that a brown bear (what we would call a grizzly bear in much of North America) had killed his cow and dragged it into a brushy ravine. We could hear his dogs barking and scrapping with the bear in the undergrowth. Ovidiu walked down into the ravine with his rifle to check out the cow and bear situation, while Christoph and I waited two hundred yards above on a ridge. BOOM! Ovidiu had jumped the bear at close range and fired a shot in the air when the bear stood up. The bear then fled, and Ovidiu fired a second round in the air after it for good measure.

The shepherds told us that they saw wolves every two or three days around their livestock, and the wolves would certainly come by now there was a rotting cow to feed on. The shepherds hated wolves and offered to help us catch them. The shepherds, and most Romanians, were not allowed to have firearms or traps, nor could they afford them if they were permitted. They were impressed by Ovidiu's rifle and were happy to help because they assumed that we'd kill all the wolves we trapped. Ovidiu confided to them that we would radio-collar every third one and let it go to track it. Just a two-thirds stretch of the truth.

We set a few traps in places likely to catch wolves but hopefully not dogs, sheep, or hungry bears. Later that afternoon, we set up our tents a half mile from the cow carcass so we could hear any commotion, be it bears or wolves. We brought lots of cheese, bread, fruit, and stinky sausage with us to last the week that we would be camped there. I started to pack my food into a backpack to pull it up into a tree with a rope so as not to draw in bears, my usual routine back in Montana. Ovidiu told me we should keep all our food in our tents with us so that the animals wouldn't get it. Hmm.

"That makes me nervous," I said.

"Don't worry," he replied. "These Romanian brown bears are not like your American bears. They are so afraid of people, they won't come into your tent. It's fine." He smiled, shrugged, and left.

I slept outside under the stars while my aromatic groceries rested in my tent fifty yards away. The brilliant full moon rose over me in the mountains in Transylvania. No vampires or werewolves yet.

During the night, we heard the bear grumbling around the car- cass and saw wolf tracks in the morning, but we caught no wolves in our traps. Within a quarter mile of our tents, in this otherwise wild valley of brown bears and wolves, we found an amazing scene: a simple shepherd's hut, crude corrals, a contented pig resting nearby, and sheep, cows, and dogs everywhere. Humans were using every square foot of the landscape to graze their livestock, collect firewood, and scavenge for food, and yet it was still the wildest landscape in Romania.

On August 9, our efforts paid off when I trapped an adult female wolf. We were ecstatic! She was a sandy gray color, in good shape, and looked quite similar to her cousin wolves in North America—which was not surprising, as gray wolves in America and wolves in Europe are all the same genus and species, *Canis lupus*, with some subspecific differentiation that morphologically affects color and size. Christoph tranquilized the wolf using a blowpipe dart that he was proficient with, and the wolf was asleep in a few minutes. We unpacked the capture gear, sample tubes, and radio collar, while Ovidiu stood guard with his rifle. We could hear three or four wolves howling at us from four hundred yards away, no doubt aware of their packmate's predicament.

We were all bent over the eighty-pound female, assessing her condition, taking vital signs, making measurements, and attaching the radio collar, when Ovidiu leapt up with his rifle and walked rapidly down the road, yelling and waving his arms. I looked up and saw a large brown bear facing us as it debated what to do next. Ovidiu was fearless as he harassed the bear, and it ran off into the forest. Wow. The two biggest Romanian carnivores in the same place, here, with us. Now my adrenaline was pumping.

This wolf was the first carnivore, and likely the first animal of any species, to be radio-collared in Romanian history. The mountain people of Romania were working hard just to survive, and research and conservation was not yet on their radar. We had a few traps stolen by shepherds, who no doubt would use them to catch and kill wolves—and so we had to recover our stolen property. When the trap thieves were chased down and threatened with their lives—or with revoking their grazing privileges, which would ruin them—missing traps miraculously appeared from clever hiding spots.

After five weeks, my time in Romania was coming to a close. I enjoyed hiking the spectacular mountains with Christoph and Ovidiu, and actually seeing wolves and bears. I heard roe deer barking at twilight, their vocalizations sounding much like rapid barks of a dog. The life of the shepherds was difficult, and I found it somehow compelling. I marveled at how they managed to graze herds of sheep for months, roaming a landscape full of predators with their only defense being their working dogs. How did they feed their large packs of dogs? They didn't pack dog food up in the mountains. One shepherd told me he fed his dogs potatoes, but I saw no evidence of potatoes in their camps, nor can hundreds of pounds of dogs survive on just potatoes. But that was as close to an answer as I got.

We visited a cheesemaking hut in the mountains, where the shepherds milked their sheep and made cheese that was ripened without refrigeration or sanitation. They sold blocks of their well-loved cheese down in the villages. They were proud of their product and graciously offered us some. Ovidiu happily consumed the samples, but Christoph and I passed. I visited Count Dracula's castle in Bran and learned about how Vlad the Impaler (also known as Vlad Dracula) ruled the region in the mid-1400s; he was a hero to his people for protecting them by cruelly impaling captured opponents in public displays. It certainly deterred people from going against him.

I was sorry to leave but was looking forward to returning to Montana, some sense of law and order, and especially the freedom to make my own choices. Even though Montana was the Wild West, it was still calmer than the "Wild East." I bid my colleagues farewell with hugs and boarded a train with my gear, including an expensive, professional-grade video camera loaned to me by a sponsor. I had taken some excellent footage of life in Romania, the wolf capture, and the beauty of the Carpathian Mountains.

I sat in my small train car with a young Romanian couple and their five-year-old boy. They spoke only Romanian, so to pass the time I drew pictures of different animals and objects and asked the boy what they were in Romanian. He would tell me the Romanian word, I would repeat it, and he would giggle in response. This was a fun game for all of us.

All was fine until we exited Romania, entering Hungary, and the train stopped. The conductor came through our car checking passports and train tickets. He pointed at my large video camera and asked for the "bond." I hadn't received any paperwork for the camera when I entered Romania, so I had no idea what he wanted. He spoke only broken English and was growing agitated as he demanded to see paperwork for the camera. The people in my car got nervous, which made me nervous.

I politely asked to talk his supervisor, hoping they would speak English and we could sort this out. He exited the train car, and my train mates looked away from me, suddenly interested in the scenery outside. I sat down in my seat, unsure what was going to happen. I was scared and envisioned spending the night in a cold, dark Romanian jail. Then the train car door flung open and a short, tight-faced female customs officer briskly approached me. She spoke English fairly well and demanded to see my "bond" for the camera. I hadn't lived in Romania for five weeks without learning a few things. Thinking about Ovidiu, I stood up facing her, my five-foot-nine frame towering over the bald spot in the center of her scalp. I produced a University of Montana business card that said I was with the wildlife department and flashed it at her. It said I was a PhD student, not faculty, but she didn't look at it closely enough to notice that.

I looked down on her and stiffly told her that I was a professor at the University of Montana and my wealthy, powerful sponsor would be very upset at this delay over the camera he had given me for my research. I firmly told her that I wanted to speak to her supervisor. She stood there silently, weighing her options. I felt sweat trickling down from my armpits over my ribs. She left abruptly. I continued standing in place, awaiting my fate. After a couple of minutes, the train started up and we began to roll along, slowly gaining speed. No customs officers returned. I looked at my train mates and they were regarding me quizzically. I made an exaggerated gesture of wiping my forehead and made the universal "phew" sound. They all smiled and relaxed. I went back to sharing fun drawings with the little boy and learning more Romanian.

IN SEPTEMBER OF 2002, I joined an Italian Alps wolf project for two weeks of the most amazing working vacation I've ever had, with fantastic playtime afterward. Francesca Marucco completed her master's and PhD degrees at the University of Montana. I was honored to be appointed to both of her graduate committees, which enabled me to join her (too briefly) in the Piedmont region of the southwestern Italian Alps to trap and radio-collar wolves, as well as ski, mountaineer, and hike.

Italian wolves evolved over centuries of persecution and were superb at avoiding humans, making them even more difficult to catch than the smartest wolves in the U.S. The wolves chose den and rendezvous sites as far from human trails as possible, which meant they raised their pups in the steepest and most inaccessible terrain they could find—and that's where we were trying to catch the adults.

Francesca is the most determined person I've ever met. Nothing deters her from achieving her ambitious goals. She and her spouse, Davide, are passionate about each other, about work, and about life. She and I instantly bonded as kindred spirits, even though she was young enough to be my daughter. They lived in the western Italian Alps, and Francesca based her research out of their mountain home—an old farmhouse-and-barn that Davide had renovated into a stunning showcase of beautiful craftsmanship and warmth.

We hiked many miles up into the rugged and spectacular Italian Alps, carefully setting traps along trails used by wolves and humans. As in the wilds of Romania, there were people in even the most remote parts of the country. After setting up our long trapline, we returned to the traps twice a day. We would run up the trails shortly before daybreak carrying heavy packs filled with capture gear and traps, to place large, flat stones over the traps to render them inoperable before humans arrived to hike the trails with dogs and children. Northern Italians are very robust and love to scramble up into the mountains to pick berries and mushrooms and picnic. With our cover-stone setting completed, we then hiked back down the mountains at sunrise to Francesca's home, where the crew would be working twenty-four-hour shifts to monitor the trap transmitters from her yard. At dusk, we would reverse the process and hike up the mountain trails to remove the cover-stones, so the traps would be active overnight to hopefully catch the nocturnal wolves. Then it was down the mountain trails again in total darkness. We were as sneaky as the best undercover agents. We absolutely could not catch a dog or have our traps discovered, or the project would get in trouble.

This was our daily routine. It was exciting, exhilarating, and exhausting. I'm a lean bean, and I still managed to lose fifteen pounds in as many days of trapping, despite enormous feasts of late-night pasta with our dedicated crew: Francesca, Davide, Luca, Tommaso, Mattia, Eglantine, my sweetheart Jonathan, and a half dozen other ambitious volunteers. Most of them were half my age, extremely fit, and fueled by their passion for life. They would work all day on nothing more than a morning cup of espresso and then gorge at night when we got home. I couldn't consume enough calories to make up for hiking twenty mountain miles a day carrying a heavy pack filled with traps, lures, and gear. I treasured the adventure and the can-do attitude and contagious enthusiasm of everyone involved. This team acted like a wolf pack—everybody respected each other, collaboratively tackled difficult tasks, managed the feast-and-famine eating routine, and pitched in to get the job done without whining. I grew to love and respect them all.

The sensitive trap transmitters sent out a distinct signal to our receivers if a trap was disturbed by humans or beasts. If the signal began beeping, we dropped everything and literally ran up the mountain with the wolf-processing equipment, drugs, and a radio collar in our backpacks to see what set off the trap. This usually happened in the black of night. Most often it was a fox, which we would release on-site, or an errant signal as the trap settled into disturbed ground. We had some wolves step on traps and spring them, and even briefly caught one wolf as the large tracks and chewed-up brush revealed—but it had managed to pull free by the time we arrived. That was so disappointing. But hope springs eternal, and every time we set off in the pitch dark, we hoped and imagined that we would find a wolf.

I've never worked so hard to catch a wolf with so many fervent helpers. Despite our extreme efforts, we had no luck during the two weeks I was there. I imagine the wolf density was about the same as my early years in the North Fork, but Italian wolves had survived centuries of persecution; only the ones that most strongly avoided people survived and passed their traits along behaviorally and genetically to their offspring. To try to lure wolves to traps set along our trail, we dragged a stinky, dead sheep across a meadow and up into the mountains. We named the sheep Dolly, after the famous sheep that was the first mammal to be cloned. We snuck around unwary hikers and mushroom pickers in the hopes they wouldn't ask us what we were doing. We howled for wolf pups to confirm rendezvous sites and rejoiced in a wolf pack's reproductive success.

After we closed and pulled the trapline, Jonathan and I explored the spectacular mountains with Francesca and Davide for the fresh scents and vistas. We saw dark blue gentians, other beautiful mountain flowers, chamois, and wild boar. But no wolves. Back at the renovated farmhouse, we enjoyed feasts of the best food I have ever eaten—and oh, the local Nebbiolo, Barolo, and Barbaresco wines! Then the four of us shared grappa (a killer liqueur) out of a five-spouted pot that we passed around to celebrate our friendship.

Francesca, however, had a thesis, and eventually a dissertation, to complete and had to collect data, test hypotheses, conduct analyses, and draw sound conclusions. She and her team eventually caught and collared a handful of wolves, but Francesca didn't have enough telemetry data to complete a graduate research project. Never one to give up, she abandoned her intended study of radio-collared wolves and instead designed a brilliant way to monitor wolf populations through non-invasive DNA sampling of wolf scats (poop) and mark-recapture methods to estimate the population size.

Mark-recapture is a method used to estimate the size of an animal population in a place where it's difficult to count individuals. The technique involves capturing animals and marking them in some way before releasing them back into the wild. In Francesca's case, instead of physically capturing wolves, she "marked" them by identifying the DNA found in scat samples. For the typical mark-recapture technique, a second independent sample is captured, the number of marked individuals is counted, and from there the population size can be extrapolated. In Francesca's case, she collected a second sample of scats. By analyzing the DNA in her second sample and comparing it with the DNA in her first sample, she knew how many of the individuals in this second sample were "recaptures." She could then estimate population size, even though she had not physically captured any wolves. It's a pretty neat technique.

She also conducted well-organized howling surveys and extensive snow-tracking surveys from skis in winter. Based on the data she and her colleagues collected, she produced a habitat suitability map that helped researchers predict wolf recolonization patterns in the Italian Alps. She continued this work into her PhD and greatly expanded our knowledge base of how to monitor wolf populations without using radio collars.

Francesca is now a professor at the University of Turin where she is the scientific coordinator of a large-scale European wolf conservation effort called the LIFE WolfAlps EU project. Francesca is paying it forward as she mentors students who are exploring the mountain trails, collecting DNA data, and connecting the wolf dots between Italy and adjacent countries. The wolf population in Italy has grown from about 100 in the 1970s to approximately 3,300 in 2021. Many of Italy's wolves have dispersed and repopulated France, Switzerland, and Slovenia—another wolf recovery success story.

MY EUROPEAN WOLF experiences gave me a new perspective on social tolerance and historical wolf-human dimensions. I had never lived in a country like Romania, where people struggled to simply feed their families and manage their crops and livestock—activities that took precedence over any thoughts of conservation. That luxury was somewhere in their future, when conditions improved. It made me appreciate my life of relative privilege. Romania has since recovered economically and become a tourist mecca, with abundant wildlife ecotours and UNESCO sites to see. I hope to revisit and experience its renewed human spirit and amazing wildlife. Christoph and Ovidiu have continued to grow the Romanian wolf project and have expanded into other Romanian conservation arenas. And the wolves are doing very well.

Since my trips to both Romania and Italy, wolf populations have greatly increased across Europe. When traditional pastoral and rural economies began declining in Europe in the 1960s, many rural residents moved away, allowing remnant wolf populations to slowly expand into the rural areas. As wolf numbers have increased, so have the number of wolf attacks on livestock, to the tune of millions of euros annually in livestock damages. The European

Commission supports the goal that farmers should receive 100 percent compensation for livestock killed by predator attacks and also be fully reimbursed for employing preventive measures to reduce attacks on their livestock. However, application and funding are controlled at the country level, often by regional authorities and local administrations. So the system is highly fragmented and varies greatly among and within European countries.

In the U.S., ranchers and farmers are also reimbursed for livestock losses confirmed due to wolves, bears, and mountain lions, but livestock losses due to predators in the U.S. are a fraction of those incurred in Europe, and compensation paid out to American livestock producers is minuscule in comparison. This is probably due in large part to the predator control work of the U.S. Fish and Wildlife Service, where problem animals, or potential problem animals, are routinely killed on the producers' behalf.

According to Dr. Luigi Boitani, a senior professor and wolf researcher at the Sapienza University of Rome, the jury is still out on whether the compensation programs and wolf removals in Europe do indeed reduce conflict or increase social tolerance for wolves. This is similar to findings in the U.S. And it is safe to say that globally, wolves are still being killed illegally, often out of frustration over perceived and real threats.

As to whether we need to reintroduce wolves in the U.S. if we want them back on the landscape, or whether it is sufficient to try to reduce conflicts and let them recover on their own, Europe offers some clues. Approximately fifteen thousand wolves now live throughout continental Europe—an increase of 1,500 percent since the 1960s. No reintroductions were necessary; wolves dispersed far and wide to neighboring countries on their own. They were joined by increasing populations of brown bears, lynx, and wolverines in rewilding human- and livestock-occupied landscapes. In some areas of low wild-prey density, European wolves eat sheep and other domesticated animals in addition to wildlife. Italian wolves were documented occasionally eating spaghetti at the garbage dumps, according to Boitani, who has studied the return of the Italian wolf for many decades. Their behavior was reminiscent of American bears feeding in dumps in the 1960s. But more recently, garbage is better controlled, wildlife populations have increased, wolves live unseen on the edges of town like ghosts, and they eat a wilder diet. What we North Americans think of as good wolf habitat— large tracts of wild habitat—simply doesn't exist in Europe. People occupy nearly all the landscapes. And yet wolves persist, and they are expanding across Europe among higher-density human populations. It's all about social tolerance. Wolf recovery is all about people and very little about wolves.

12

·····

LIONS AND WOLVES
AND BEARS, OH MY!

N 1978, THE YEAR before my adventures started in Moose City, I
was looking for red foxes in the lake country of northern
Minnesota. A wizened old wolf trapper told me, "To find fox
you look for wolf tracks, cuz them wolves and fox get along okay.
But if you see coyote tracks, hell, you might as well forget about
finding fox, cuz coyotes will kill a fox every chance they get." The
leathery-faced woodsman spoke fondly of many years spent out
on his traplines. He may not have had a formal science education,
but he had a strong grasp on the concept of ecological trophic cascades, which means the ways animals at the top of a food chain
affect each other and, in turn, the animals and plants further down
the food chain.

I witnessed this dynamic process in the wilds of northwestern
Montana as populations of predator and prey species significantly
changed when wolves recolonized the area. In my early years
of snow tracking our first wolf colonizer on the North Fork,
Kishinena, I saw coyote tracks everywhere, I rarely saw wolf tracks,
and I never saw fox tracks. Then I remembered that old Minnesota wolf trapper's wisdom. After the eradication of wolves from
the Rocky Mountain west, coyotes filled in the space, or niche,
that wolves had occupied. As coyote numbers increased, red fox
numbers decreased due to competition from coyotes. Eventually,

as wolf numbers rebounded, coyote numbers decreased and fox numbers rebounded. With each step up the food chain, from a ten-pound fox to a thirty-pound coyote to a ninety-pound wolf, there appears to be an ecological equation where a threefold size difference triggers competitive exclusion for food resources. What this means is that wolves kill coyotes and coyotes kill foxes, because their prey preferences overlap. However, wolves and foxes eat different things (called niche separation) so wolves better tolerate foxes.

WITH THE RETURN of wolves, the complex landscape of the North Fork and Glacier National Park area supported nine medium-to-large predator species (red fox, coyote, bobcat, lynx, wolverine, wolf, mountain lion, black bear, grizzly bear) and four big prey species (white-tailed deer, mule deer, elk, moose) that most of these predators competed for. By the early 1990s, after wolf populations built up in the Flathead, my friend and colleague Wendy Arjo was doing her PhD on coyotes. She documented a decrease in coyote density as well as a shift of coyotes to safe zones, called buffer zones, where they lived between the edges of wolf pack territories, in which the wolves spent less time. And with fewer coyotes around, we started seeing fox tracks. The canid balance shifted now that wolves were top dog. Fox dens began turning up in embankments, under sheds, in culverts, and near campgrounds. We can thank the wolves for the return of Reynard.

Several years after wolf reintroduction to Yellowstone in 1995, a similar wolf-coyote scenario was documented. Bob Crabtree had been studying coyotes in Yellowstone for years prior to wolf reintroduction. As wolves became established, coyote numbers decreased and coyotes began using the landscape differently to avoid wolves—and foxes were more commonly seen. Déjà vu to our Glacier Park work.

Trophic cascades became a popular topic; the scientific theory has remained hotly debated in the scientific literature from the

2000s to the present. The debate is not whether it occurs (it does, as noted by the wolf-coyote-fox examples), but how great or small the effect is and how far through the system it can be tracked by research. Various researchers studying the Yellowstone ecosystem have published a wide range of conclusions about trophic cascades—ranging from wolf recovery having little effect to wolves influencing the ecosystem far down the food and plant chain.

A video by British environmental journalist George Monbiot called "How Wolves Change Rivers" has had more than 44 million views online. The video claims that wolves move the elk herds around to such a degree that they reduce their browsing on aspen and willow. In turn, the video continues, river bottom aspen and willow regenerated after being released from fifty years of overbrowsing by elk, which in turn allowed beaver to return, build dams, and create wetlands, bringing back neotropical songbirds, fish, and amphibians, and even changing the behavior of the rivers themselves. Really big stuff—but is it actually true?

The reality is much more complex than this. Are wolves to blame for the reduction in elk numbers in Yellowstone National Park? I have often heard that claim from anti-wolf people. Researchers, however, have shown that elk populations are affected by a combination of climate change, drought, fire, winter severity, grizzly bears, black bears, mountain lions, coyotes, and human hunters, in addition to wolves. Sorting out which component has had the most impact on elk has not yet been accomplished and likely never will be; relationships between weather, plant productivity, animal nutrition, carnivore and elk use of space, and competition between several species of animals are intricately entwined.

THE FAINT BEEPS from the wolf's radio collar coming through my headset were growing stronger over the incessant roar of the Cessna's engine as we homed in on the wolves. I wiggled my numbing toes inside my winter boots to warm them up as we cruised above

Glacier National Park on this subzero February day. Flying in that overpowered, unheated dragonfly with pilot Dave Hoerner was magical; we had shared hundreds of hours together in the sky searching for wolves. After hearing a series of faint beeps from an altitude of nine thousand feet, we spiraled down four thousand feet in elevation while we flipped a switch back and forth between the antenna on the left wing to the antenna on the right wing.

This switching between the antennas mounted below each wing of the plane enabled us to keep the radio-collar signal within our shrinking flight circle as we worked to pinpoint the location of the radio-collared wolves. There was no room for error while we were circling at treetop level in mountainous terrain at a speed just shy of stalling the plane, all the while trying to count wolves weaving their way through the forest below. Dave was a superb pilot and I trusted him with my life every time we flew in the mountains.

I slowly turned down the volume on the receiver to home in on the exact location of the wolves. At the same time, Dave and I scanned for animals. Deep snow shrouded the ground, making it a little easier to see dark shapes moving through the tall trees. The radio-signal beeps became nearly deafening as we flew directly over the wolves and my gloved hands fumbled with the knob to turn down the volume yet again. Eventually, Dave spotted the roiling mass of gray and black wolf bodies, and he banked the plane in a tight circle above the melee.

The stall alarm blared intermittently as we tried to determine what was going on below us. The wolves had just killed a large animal and were tugging on the carcass. The victim's compact, tawny shape didn't match my impression of a leggy deer or elk. And then it clicked—a mountain lion! The pack had just killed a large mountain lion and the wolves were reveling in the spoils of vanquishing a competitor. As we continued circling above this remarkable sight, the wolves started to lose interest and began wandering off. The last one to leave the kill site was the large dominant male, who walked over to the lion, cocked his hind leg, and urinated on the mountain lion's head.

I badly wanted to examine the kill site, but it was across the river and many miles from a road. As Dave maneuvered the plane to a higher altitude to gain more landscape in our view, the two of us pored over paper maps spread across my lap to determine our location in the featureless backcountry. We agreed that we were somewhere in the Anaconda Creek drainage in Glacier National Park, but we were miles from any recognizable landmark in the heavily treed landscape of the Flathead Valley. After thoroughly memorizing the landscape pattern, we flew the thirty-five miles back to Moose City and landed the little plane-on-skis on the crude, snow-packed runway in the meadow.

Dave taxied the plane closer to my cabin so I could run in and get some equipment to use as a location system: a wolf radio collar, bubble wrap, a cardboard box, and duct tape. I climbed back into the plane, and we took off, heading south toward Anaconda Creek, while I turned on the radio collar and tested it, packed it inside the box in a beehive of bubble wrap, and wrapped a seemingly endless amount of duct tape around the outside of the box. The wolves were a quarter mile away from the dead mountain lion and curled up resting when we flew back over the kill site. I pushed the plane door open about a foot, and when Dave had perfectly placed us in a tight circle over the mountain lion's body, I leaned out into the freezing wind and pitched the box out, watching it disappear into the snow fifty feet north of the lion. Bingo. Location system deployed! As we flew off, I listened as the boxed beeps came in reassuringly strong from below, and I felt good about being able to return to get a look at the death scene.

The next day, WEP volunteers Gray Neale and David Pilliod pulled on chest waders, waded the icy North Fork of the Flathead River, skied two miles in, and stayed overnight in the Logging Creek backcountry cabin in Glacier Park. The following day, they skied many rough miles through entangled deadfall, carefully threading between the trees with the delicate antenna—an essential piece of equipment to locate the radio signal—strapped onto the outside of a backpack. Periodically, they used the antenna and receiver to check on the locator beacon signal to make sure they were heading in the right direction, eventually coming upon the kill site and the dead mountain lion.

The grim kill scenario was easily reconstructed through tracks and blood in the snow and on the tree trunk. The wolves had surprised the mountain lion and it had bolted up the nearest tree to escape the nine rushing wolves. Unfortunately for the mountain lion, that tree was a spindly lodgepole pine. The cat climbed up as high as it could, but the tree had no large branches for it to securely perch on for the time needed to outwait the wolves below. The cat's claws had left vertical grooves in the tree's bark as the doomed mountain lion slid down the trunk into the waiting pack of wolves. The ferocious battle was over quickly once the cat hit the ground. The wolves pulled out some of the intestines, but they didn't feed on the carcass. Like most carnivore-on-carnivore kills, this was simply an act of eliminating a competitor and not a hunt for food.

WOLVES HAD ONLY recently returned to the North Fork and mountain lions were having to learn rapidly how to avoid them. Since wolves and mountain lions hunted basically the same prey species across the same large landscape, they began to separate spatially. It is the evolutionary outcome of stalking (mountain lions) versus coursing (wolves) predators. A stalking predator uses cover to ambush prey and its chases are generally a short sprint, whereas a coursing predator prefers more open terrain to chase and run down its prey over longer distances. So generally, mountain lions live in rougher terrain with more forest cover for stalking prey—and also more cover if they need to escape wolves. Wolves choose less rugged, more open habitat where they increase their success of detecting and chasing down their prey. Wolves are built for long endurance chases (think marathon runners), whereas mountain lions are built for short sprints and leaping with more compact, muscular bodies (think wrestlers). I had captured and handled dozens of wolves before I accidentally trapped a mountain lion; I was amazed to feel the thickness and power of its muscular legs compared to the sinewy running legs of wolves.

But the mountain lion may have an interesting advantage over wolves. In 2022, researchers in Yellowstone suggested that a parasite found in wolves, *Toxoplasma gondii*, makes the wolves bolder, more likely to engage in high-risk behavior, and forty-six times more likely to become pack leaders. Infected wolves are also eleven times more likely to leave their packs and go off on their own. All these behaviors increase the chances of death for both the infected wolf and for the members of its pack. Once infected, a wolf hosts the parasite for the rest of its life. But as it turns out, the wolf is an intermediate host for the parasite, which means that for the parasite to complete its life cycle, the wolf must die so the parasite is freed to be picked up by its primary host, a mountain lion.

The interesting part is that cats, and mountain lions in particular, are relatively unaffected by infections of *Toxoplasma*. This parasite must live in a cat to reproduce sexually and complete its life cycle, after which it is shed out in the feces of the mountain lions. When wolves steal mountain lion kills, wolves could become infected by eating the mountain lion or by ingesting its feces. Then this microscopic parasite may manipulate the behavior of wolves, turning them into risk-taking individuals with higher death rates, which benefits mountain lions in the long run. This parasite exchange may go back more than twelve thousand years, to a time when American lions the size of modern African lions successfully killed infected wolves. Turnabout is fair play.

Toni Ruth of the Hornocker Wildlife Institute studied mountain lions in the Glacier area and documented numerous lion-wolf interactions. As a mountain lion is generally larger and more powerful than a wolf in a one-on-one encounter, a mountain lion will win the battle with its sabered paws, rotating forearms (like a human's), and strong jaws. But a pack of wolves will overpower a mountain lion if it cannot escape up a tree—where it can perch in the limbs—or leap onto a rocky ledge quickly enough. Thousands of years of coevolution is the reason that a 130-pound mountain lion can be treed by a 40-pound baying hound dog.

Up the North Fork, both wolves and grizzlies drive mountain lions from their kills. A mountain lion is a superb killing machine, which is why they don't need to live in packs. But Toni found that for a young, inexperienced mountain lion, a mother with kittens to feed, or an injured mountain lion that has to repeatedly give up its kill to wolves or bears and then hunt again, it can mean starvation for the mountain lion, or sometimes immediate death as a result of these direct encounters.

In the North Fork, there was a third large predator battling wolves and mountain lions for food—the grizzly bear. It became a numbers game for which animals had the most mass, attitude, and fortitude at a kill to determine who would possess the spoils. A pack of wolves may or may not successfully defend a kill against a large grizzly.

One October I watched, from the safety of our circling Cessna, as a large male grizzly defended the carcass of a bull elk against seven hungry wolves. I don't know if the wolves had originally killed the elk; or if it had been wounded by hunters, gotten away, and died; or if possibly the grizzly killed it. The grizzly lay on top of the elk, looking menacingly at the circling wolves, no doubt vocalizing its displeasure with deep throaty rumblings. For their part, the wolves circled excitedly, tails high and waving, feinting small lunges toward the elk.

One bold gray wolf suddenly dashed in and grabbed at the elk, which provoked the bear into charging and chasing the daring wolf into the trees. As soon as the bear was preoccupied with chasing the thieving wolf, the other six wolves all bounded to the elk carcass—and savagely tore off and gulped down hunks of meat. Suddenly, the bear stopped as the light bulb went off in his bruin brain, and he looked back at the elk and saw that he had been duped.

The grizzly charged back to the elk, the wolves scattered, and the bear settled in on top of the elk again. The wolves circled and taunted the grizzly, and then a black wolf crept in to sneak a bite of elk steak. The enraged bear launched off the elk and chased the black wolf into the woods as it ran just out of reach. Once again, the remaining wolves rushed in and swarmed the elk, voraciously ripping off pieces of flesh and gulping them down until the bear roared back to guard his banquet. I saw this happen three times, involving at least two different wolves, which made me wonder if this was a tried-and-true strategy for this wolf pack, or for all wolf packs? I don't know who eventually ended up with more meat, the wolves or the grizzly, but I'd put my money on the wily wolves.

TRACKING WOLVES ON skis in the North Fork gave us opportunities to read these epic carnivore encounters more easily in the snow. Through a dozen winters, keen field observers on the WEP crew, the mountain lion research team, and the grizzly group found hundreds of kills and pieced together a fascinating account of who was who in the carnivore world. I was amazed to see grizzly bear tracks every month of the winter. I had naively presumed these bears would be comfortably sleeping in their dens and I would be safe skiing along wolf trails in three feet of snow.

I often wondered if the additional bonanza of wolf-kill carrion caused a few grizzlies to give up denning in favor of the food rewards during the cold season. Did grizzlies stay out all winter before wolves recolonized the area? Was it all bears adopting this strategy or just a few? The wandering, winter grizzly tracks we observed were always from large single males—some of the prints were as long as my eleven-inch ski boot—and we never saw tracks of young bears or family groups in the snow. I thought it would be fascinating to radio-collar these winter bears and find out more about them. Were they old, starving, sick, healthy, cantankerous, or adventurous—the same individuals or different bears each winter?

I KNEW THAT the Camas Pack (formerly part of the Magic Pack) had made a kill near Hidden Meadow because of radio-tracking signals from the North Fork Road; their collar signals didn't move for three days, and then the wolves left the area. Wolves are relentlessly restless in their hunting efforts and territory maintenance to repel any trespassing wolves. Territory marking to convey "this land is taken" is a daily activity, along with finding food to support the pack. Wolves don't stay put unless they have a big pile of meat to feed on. So, on this cold January day, I waded the river in chest waders, the ice chunks bumping into my legs as the river flowed into and around me. Using my ski poles for balance, I carefully navigated over ice-covered rocks and through the moving current.

I skied four miles, into the kill site, only to be confronted by a blowdown of spruce and lodgepole pine. I am a skilled obstacle skier, but this tangle of downed trees was so dense that eventually I had to take off my skis and wade through thigh-deep snow. It was a painstakingly slow process of putting my weight onto my leading foot, transferring my weight, and then punching down through the crust—what we called post-holing. Very slowly, I rose up and sank down, slogging my way along the wolf trail, when ravens suddenly burst up off the ground seventy-five yards ahead of me, squawking and croaking as they noisily flapped away through the trees. That riotous outburst startled me, so I paused to collect myself before post-holing the last few yards to the kill. About halfway there, I vaguely saw a large, brown animal get up and disappear into the dense brush. Probably a moose; I'd had the bejesus scared out of me by a moose two days earlier and I wasn't going to be fooled again. I finally came upon the well-gleaned remains of a bull elk with a nice five-by-six-point set of antlers. Data!

I pulled out my bone saw to cut open the femur so I could visually examine the fat content of the marrow. This visual field method is a quick way to assess the health condition of elk. White, fatty marrow like a soup bone meant the elk was relatively healthy, while red and gelatinous or almost-clear marrow, approaching the look of apple jelly, meant the elk was near death from starvation, as it had burned up the last of its fat reserves. I began jotting notes with a pencil in my waterproof field notebook. Then I went to work with my knife to skin back the carcass, making it easier to look for bite marks and wounds to determine how the animal had died. I filled sample bags and assessed how the large bull had been pursued and killed, and by what carnivore, confirming it had been killed by wolves.

Then curiosity got the better of me, so I waded through snow splashed with raven poop to where the moose I'd seen running away had been bedded, only to discover that what I'd thought was a moose was actually a grizzly bear. Gulp. The bear had been resting next to the dead elk while feeding on it. I sucked in my breath, my pulse quickened, and I looked around to see if Mr. Griz was waiting to pounce on me. Nope.

I went back to detailing how far the wolves chased the elk, how old the elk was, and how many wolves there were. The femur marrow was red and runny, so the elk was starving. The carcass had been 95 percent consumed by the wolves, bear, and ravens. The fact that there was little food value left on the carcass was my saving grace and the reason I had not been charged by a grizzly intent on defending its meal. I packed up my gear and samples and slowly struggled my way back to where I had left my skis. A chill ran through me as I saw fresh grizzly tracks that crossed on top of my ski tracks. The bear had circled me and examined my tracks but left me alone. He could have had me for dessert. Fortunately, he chose not to. That was typical of field research and life among grizzlies, despite the terrible tales people repeatedly told about crazed bears.

THERE ARE THREE main ways that competing carnivores, including the diverse predator guild in the North Fork, minimize conflicts. They select different prey species, like foxes and wolves do, to lessen food overlap. They occupy different habitats in the same landscape, like wolves and mountain lions do, to separate themselves spatially. Or they use the same habitat but at different times, like coyotes and wolves do, to distance themselves temporally. All three tactics create what are known as habitat refuges—places of safety through separations in food, space, and time. Habitat refuges—along with other behaviors, like supporting offspring for longer so they can grow more before heading out to survive on their own—enable competitors to coexist over the long run. For mountain lions, for example, this gives the kittens a protective mother for longer to help ensure their survival. Mother Nature has devised many methods of survival for all her creatures.

In all those North Fork years in the 1980s and 1990s, we found no evidence of wolves interacting with bobcats or lynx, nor did I find evidence of that in the scientific literature in North America. Perhaps since bobcats and lynx select smaller prey, wolves do not see them as competitors and therefore disregard them. Bobcats and lynx may even benefit from scavenging on wolf kills. Although it was uncommon to find evidence of run-ins between the top three carnivores in the North Fork, particularly during the summer with no snow to show the story, tracing what happened when they met was one of the most enthralling aspects of our research.

As wolves continue to expand their range across North America, I am hoping that this fascinating world of carnivore dynamics will continue to be investigated and further understood by humans. Different carnivores have evolved ways to coexist or eliminate their competition, and they will defend their home range to the death. They really aren't so different from us.

13

SLAYING
THE SUPER-WOLF

T HE WOLVES THAT slipped into Montana's wilder landscapes under their own power in the 1980s settled into forested areas and were mostly accepted as a natural part of the wildlife scene. They were shy of people and not particularly controversial, although hunters and ranchers bellyached some. Wolf proponents thrilled at the sight of tracks and rejoiced to hear melancholy howls drifting through the pines. By 1990, a groundswell of public wolf love was emerging across the country. Many nonprofit organizations were doing wolf outreach and education, and working with congressional staff to build support for wolf reintroduction. Wolfy souvenirs, books, and housewares began showing up in stores everywhere.

When conversations began in earnest about wolf reintroduction to Yellowstone and Idaho, I was reluctant to support it because the WEP had documented wolves successfully recolonizing Montana—and a handful were making it to Idaho and Wyoming without human assistance.

Take wolf 9013, a 110-pound male that I radio-collared in Glacier National Park, who drifted over to Idaho and set up his territory in the Kelly Creek area in 1991, where he survived as a lone wolf for several years. Four years later, a reintroduced black female wolf dispersed into the Kelly Creek area and paired with 9013. There, they raised many pups over the years until 9013 finally died of old age. Meanwhile, in the Yellowstone area in August 1992, Bozeman cinematographer Ray Paunovich filmed an adult black wolf with silver grizzling on his fur comfortably feeding on a bison carcass alongside coyotes and a grizzly bear in Yellowstone's Hayden Valley. Wolf lovers, wolf haters, and scientists debated whether this was truly a wild wolf, a captive animal that had escaped, or a pet wolf-dog turned loose in the park. Many of us who watched Paunovich's footage at the North American Wolf Symposium in Edmonton, Alberta, in 1992 agreed that the animal looked and acted like a typical black wolf from the Glacier National Park area. In September 1992, a hunter illegally shot a ninety-eight-pound, all-black, male wolf in the Teton Wilderness immediately south of Yellowstone National Park. This wolf was not the same one that Paunovich filmed. The federal lab in Ashland, Oregon, analyzed its DNA and determined that it was genetically most similar to wolves in the Ninemile Pack near Missoula, Montana. Clearly, this wolf had successfully walked the long journey from Missoula to this area in Wyoming, passing through Yellowstone.

The longest dispersal the WEP documented was a black female, two-year-old wolf 8551, who ran north for a straight-line distance of 540 miles to Pouce Coupe, British Columbia, where she was shot in July 1987 while traveling with three other wolves. Mike Fairchild had radio-collared 8551 as a pup and I had recaptured her and fitted her with a new radio collar when she was a year and a half old; both captures were within ten miles of Glacier National Park. The Canadian shooter gave 8551's radio collar to his local game warden, who called the phone number that I had written inside the collar with a permanent marker. The game warden talked with Bob Ream in Missoula and gave Bob the details of her end point; they both marveled at the distance she had traveled. I was sorry to hear that she was dead, but what mind-boggling data that black wolf provided about the role of dispersal in wolf population connectivity!

If she had traveled south instead of north, she would have ended up about one hundred miles south of Yellowstone National Park. Clearly, wolves were leaking out of the native recolonizing population in northwestern Montana and setting up territories far away. Wolf recovery through natural dispersal was working, albeit slowly. They just needed more time to reach critical mass.

IT IS HUMAN NATURE to resist things that are forced upon us—be they religion, politics, or wolves. I absolutely believed that wolves would disperse to Yellowstone, Idaho, Washington, Wyoming, and beyond on their own four feet, and in doing so would be protected as a fully endangered species under the Endangered Species Act. That was my preferred method for them to recolonize the spaces where they had once roamed freely, because wolves that were reintroduced, rather than making the journey on their own, would not be protected as endangered, but instead would be labeled as a "non-essential experimental population," which would allow sanctioned killing and management flexibility anywhere outside of a national park. The critical question was whether it would take them ten years or a century to return.

In the end, the question became moot as the powers that be decided to forge ahead with wolf reintroduction, and three wolf management zones were created: Northwest Montana Recovery Area (a zone north of I-90 that included part of northern Idaho), Central Idaho Experimental Area (which included parts of Idaho and Montana), and the Greater Yellowstone Experimental Area (which included parts of Montana, Idaho, and Wyoming).

The wolf reintroduction came about after a highly controversial and somewhat surprising collaboration involving wolf proponents and wolf opponents. Vocal wolf enthusiasts were pushing wolf reintroduction, and Democratic representative Wayne Owens of Utah introduced legislation to Congress in 1987 to reintroduce wolves to Yellowstone. But also in the background were some wise, conservative congressmen representing ranchers who hated and feared what wolves would do to their livelihood. Wolves came to symbolize ranching economic woes, young family members leaving the ranch for a different lifestyle, private property rights, big government, and many other forces that ranchers felt were threatening their values. Wolves were a flesh-and-blood scapegoat and easily targeted.

Republican senator Jim McClure of Idaho, who was no wolf lover, saw the writing on the wall as fully endangered—and therefore protected—wolves began showing up in unexpected places. McClure, one of the most senior and powerful people in Congress, backed wolf reintroduction, but only with added terms and conditions that would protect ranchers and his conservative Idaho citizens. Other conservative Republican congressmen also reluctantly saw the value in joining the wolf reintroduction movement so they wouldn't have to deal with an unwanted endangered species in their state. In July 1993, conservative Republican senator Alan Simpson of Wyoming was quoted in the *Casper Star-Tribune*: "If we're going to have it shoved down our throats, it should be done as an experimental population so we have the proper management flexibility."

Federal wolf reintroduction was slated for central Idaho and Yellowstone National Park. The plan was that a group of U.S. and Canadian trappers, pilots, and biologists would capture wolves in Hinton, Alberta, and Fort St. John, British Columbia, for reintroduction to Yellowstone and Idaho in 1995. But there was the rub. What kind of wolves would these be, exactly? Would they be the kind of wolves that had originally lived in the Lower 48? Or would they be some type of foreign wolf that had no business being here at all? And here is where the WEP research came into play. Eight years earlier, wolf 8551 had trotted north past Hinton and settled in the area around Fort St. John. Plucky 8551 showed us that there was indeed one native wolf population from Yellowstone to the

Yukon, and the journey between these two locations was merely a long walkabout for a wolf.

Although 8551's trek had been an extraordinary feat, we did not realize how important her journey was until the anti-wolf-reintroduction crowd began screaming about the foreign Canadian super-wolves that the federal government had moved from Canada and dumped into Yellowstone and Idaho. Never mind that the reintroduced wolves looked, acted, sounded, and smelled like the wolves that had walked into the North Fork and northwestern Montana from Canada. Cries went out that these reintroduced super-wolves weighed 175 pounds and had eight canine teeth—and they were very dangerous because the most aggressive wolves had been selected for reintroduction, so they would survive no matter what. No one cared that the weights and body shapes of the wolves in Montana, Idaho, and Wyoming were comparable. Not a 175-pounder in sight. Not even a 150-pounder.

FOR ITS PART, the Idaho legislature barred the Idaho Fish and Game department from involvement with the 1995 reintroduction. At that point, the Nez Perce Tribe approached the federal government and requested to take the place of the state. The U.S. Fish and Wildlife Service accepted the tribe's wolf recovery and management plan, making the Nez Perce the first tribe to reintroduce wolves and oversee the statewide recovery of wolves.

A total of seventy-six wolves (sixty-six wolves captured in Canada, plus another ten wolves from the Sawtooth Pack along the Rocky Mountain Front in Montana) were reintroduced to Yellowstone and the central Idaho wilderness from 1995 to 1997. The wolves in Yellowstone turned out to be highly visible in the wide-open landscapes of the Lamar Valley, and wolf proponents, wolf watchers, and researchers were given the chance of a lifetime to observe wild wolves in their natural habitat. What an incredible opportunity for all.

Thousands of articles and hundreds of books have been written about the Yellowstone wolves. Behaviors and interactions never before seen were documented, to the enrichment of wolf science and ecology in general. Love of wolves boosted ecotourism to more than $35 million, pumped annually into the economy of small towns on the edge of Yellowstone as tourists flocked to see wolves, buy wolf mugs and art, eat meals, take ecotours, and stay in motels. Wolf proponents rejoiced as their dreams came true. This was in stark contrast to the growing number of vocal and well-funded anti-wolf voices in the hunting and livestock industries. The wolf love-hate divide deepened and cascaded through sociopolitics, like a freight train picking up speed with each passing mile.

IN THE MEANTIME, the wolves did what wolves do. They reproduced successfully and the three separate populations in northwestern Montana, Yellowstone, and central Idaho eventually merged through dispersal to form a single metapopulation. Dispersers from this population then set up new packs in Montana, Wyoming, Idaho, Washington, Oregon, California, and Colorado. The wolf was officially back in the western states.

Eventually, the official wolf recovery goals were met, and reintroduction was considered so successful that plans were laid for the wolf to no longer be considered an endangered species under the federal Endangered Species Act in three states in the Lower 48: Idaho, Montana, and Wyoming. It was now up to each of these three states to present a wolf management plan to the federal government to ensure the survival of the species in each state in the absence of federal protection. Wolf recovery plans for Montana and Idaho were accepted by the U.S. Fish and Wildlife Service and wolves were delisted in these two states in 2011. Wyoming, however, wouldn't play ball and wrote a wolf management plan that was unacceptable to the U.S. Fish and Wildlife Service. Wyoming's plan was rejected and the wolf's status in that state was in and out of the court system, delisted and relisted several times, including in 2006, 2009, 2012, and 2014. Wyoming eventually wrote an acceptable plan and wolves were finally delisted there in 2017.

THE POLITICALLY DIFFICULT times brought on by such a successful wolf recovery cast a shadow over the latter part of my career. In 2016, I was hired as the wolf and carnivore specialist with Montana Fish, Wildlife & Parks. By that time, the disgruntled rhetoric had taken on a more fevered, violent pitch, and state employees were easy targets. Wolf madness had entered the halls of state legislatures, governors' offices, and federal administrations.

In Montana, Idaho, Wyoming, and other western states, the majority of funding for the state wildlife management agencies comes from the states' sales of hunting, trapping, and fishing licenses and federal excise taxes paid on purchases of firearms, ammunition, and fishing equipment. A large percentage of hunters who target deer, elk, and moose seem convinced—despite field survey data to the contrary—that the wolves have greatly reduced big-game populations and are competing with hunters. Trappers back them, so they want wolves gone. Hunters and trappers have been stridently vocal about their anti-wolf sentiments, and state wildlife managers and legislators are listening. Western state policymakers and game department officials have proven to be more concerned with accommodating the enmity of big-game hunters and fur trappers toward wolves than with maintaining a natural balance of predators and prey based on sound science. Thus, the radical legislation to decrease wolf populations.

Wolf populations in the Lower 48 had been downlisted, delisted, and relisted several times, sometimes with wolves protected as endangered or threatened, and sometimes with protections totally removed. The wolves were unfairly being made pawns in this game of Whac-a-Mole, and a lot of dead wolves were piling up.

I OFTEN HEARD about the foreign super-wolves—somebody who knew someone's brother-in-law who had been attacked by a wolf. There were tales of the alleged annihilation of deer and elk populations due to these monstrous killers. A biologist friend of mine was asked by one fellow if they knew who was really behind the wolf reintroduction. When she wisely declined to answer, the man knowingly leaned forward and told her that it was insurance companies. "Oh?" she said, not seeing the relevance. "Yeah, the insurance companies pay out so many millions every year in deer collision damages that they wanted the wolves to thin out the deer herd to save some money." Another claimed, "My uncle saw a government truck dumping out wolves in the Big Snowy Mountains. And you government folks tell us there are no wolves here. Bahhh." Another fellow told me that he had seen government employees on snowmobiles dump wolves out of crates in the forest north of Kalispell in 2006. I know for a fact that this did not happen, but I decided to keep my mouth shut because arguing with him wouldn't do any good and would just make him a more ardent wolf-hater.

SOMETIMES IT WAS hard to even remember the early years, when people actually enjoyed seeing a wolf and were thrilled to hear wolves howling. These days, when I need to recharge my batteries, I take a road trip to Yellowstone and join the scores of wolf watchers with their scopes on the side of the road, admiring the animals they love. The wolf watchers and researchers are so enthusiastic about seeing a wolf pack through powerful scopes—hunting elk or sleeping on the snow a mile away. It nourishes my soul. I go up to my cabin near Glacier National Park and, if I'm lucky, I find wolf tracks and may follow them on my skis for a while. It refreshes many precious memories of wolves and the joy of discovery in the early wolf recovery days.

As of this writing, approximately 3,000 wolves roam the six western states. This population remains linked to another 15,000 wolves in British Columbia and Alberta. Another 4,500 wolves live in the Great Lakes states of Minnesota, Wisconsin, and Michigan, and are also connected to a larger Canadian population. And another 10,000 wolves roam the wildlands of Alaska, where they have never been threatened or endangered. I cannot think of a more profoundly successful endangered species recovery story—or a more controversial one. Wolves are loathed and feared, as well as respected and loved, depending on your set of values. Almost no one is neutral about wolves. Dave Mech, the world's foremost wolf expert, said it best in an article about the wolf: "It is neither saint nor sinner except to those who want to make it so."

14

FEAR AND WOLVES

"LISTEN TO THEM—the children of the night. What music they make!" mused Count Dracula as the wolves howled in the valley below his Transylvanian castle in the classic 1897 novel *Dracula* by Bram Stoker. I still remember being terrified of Dracula, played by Bela Lugosi with fangs and flowing black cape, in the 1931 black-and-white horror film based on the book. Although I don't remember the wolves howling in the film, the howl of the wolf has spawned fear in humans that has passed along through myths of wolves that devour the sun, guard the underworld, and eat grandmothers. Perhaps worse yet were the cursed human werewolves (lycanthropes) who morphed into vicious wolves when the moon was full and lived undetected among humans the rest of the time, shape-shifting between the two worlds and belonging to neither. Lon Chaney Jr. played the role of the werewolf in the 1941 movie classic *The Wolf Man*. In the movie, when werewolves were mentioned, the villagers recited a poem: "Even a man who is pure in heart and says his prayers by night, may become a wolf when the wolfsbane blooms and the autumn moon is bright."

In more recent years, there have been rock songs, soap operas, television series, and video games that continue to keep this myth alive. I'd like to think that those who follow them do so in jest, but the very fact that this genre exists indicates otherwise.

The graphic documentary film *Death of a Legend*, produced by the National Film Board of Canada and released in 1971, was the first wolf movie I watched that was firmly on the side of the wolves. I still vividly remember scenes from the exceptional footage of wild wolves hunting, living, and dying. The movie explained the ecological role of the wolf in maintaining the balance of nature and tried to dispel the image of wolves as malicious beasts.

I HAVE LONG pondered why people fear wolves so much. Every year, there are roughly 4.7 million humans bitten by dogs worldwide. Approximately thirty thousand of these bites are fatal, with an average of forty fatalities in the U.S. annually. In the past one hundred years, a total of 126 people were attacked by wild mountain lions, of whom 27 died, with attacks increasing significantly in frequency in recent years. From 2000 to 2020, twenty-five people were killed by wild black bears across North America, roughly one attack per year. In the last twenty years, approximately thirty to forty people were killed by grizzlies in North America. There are far fewer grizzlies than black bears, with grizzly populations only in western Canada, Alaska, Montana, and Wyoming, whereas black bears live in nearly every state and province in North America.

In North America, approximately two dozen attacks (mostly non-fatal) of people by wild wolves have been documented in the past one hundred years. Some of these people were bitten by rabid wolves and eventually died from rabies rather than from the initial bite. Only two human fatalities have been documented in North America in the past twenty-five years: wild wolves killed Kenton Carnegie in Canada in 2005 and Candice Berner in Alaska in 2010. Like with all carnivores, the number of attacks has been increasing more recently as people and carnivores overlap more in time and space. The chances of being attacked by a wild wolf are far too low to calculate, but they are a little above zero. Here's another way of looking at the statistics. Cows kill roughly twenty people per year in just the U.S. Roughly thirty people die from lightning strikes in the U.S. each year. Approximately two people are killed annually in the U.S. after tipping over a vending machine that then crushes them. Wolves? Much, much less frequent offenders than all these others. And yet people fear wolves almost pathologically and don't harbor this fear against bears, mountain lions, cows, or vending machines.

Some people believe that the 1918 novel *My Ántonia* by Willa Cather is a factual account of a real European wolf tragedy. Pavel and Peter were driving a wedding party to the bride's home and saved their own lives by throwing the bride and groom from their sleigh into a pack of attacking wolves. Similar accounts appear in the mythology of Russia and Europe, but none have been verified.

With our current advances in technology, extraordinary ecological research, DNA discoveries, and instant information, people have the tools to understand the science about what wolves are and, perhaps more importantly, what wolves are not. But cultural myths are often more powerful than knowledge, and we are not a lot further advanced in our present wolf perceptions, despite the high volume of good-quality wolf information that is disseminated daily. With a tiny DNA sample from tissue, scat, or fur, a researcher can determine the population source the wolf came from, which is important for understanding the connectivity between populations; if there is dog or coyote somewhere in the wolf's lineage; the gender, parentage, level of inbreeding, disease resilience of the animal; and much more.

To take a specific example, advances in the field of molecular genetics, based on DNA, are growing rapidly. In the mid-1990s, I wanted to look for genetic relationships that coded for black coat color, using the accurate pedigrees the WEP had created from field observations of known North Fork wolves.

We had the tissue samples frozen, awaiting future uses such as genetic analyses. Dr. Fred Allendorf, renowned conservation geneticist, was on my PhD committee, and he said that the molecular techniques were not yet refined enough to assess coat color relationships using genetic analyses. However, recent genetic research in Yellowstone has allowed researchers to determine that black coats in wolves were due to a gene mutation, the K locus gene, that was likely derived from interbreeding with Eurasian domestic dogs brought to North America 7,000 to 14,000 years ago. Dogs developed this mutation through the domestication process as they split from wolves 11,000 to 34,000 years ago. The black coat color is a dominant mutation of a single gene. Therefore, a black wolf can be homozygous (KK) or heterozygous (Kk), and their pelages are indistinguishable, but their fitness is very different. Heterozygous black wolves live longer and reproduce better than gray wolves, and gray wolves fare better than homozygous black wolves.

Further comparisons in Yellowstone revealed that coat color is more than just a pretty nuance. Wolves with black coats apparently survive canine distemper virus (CDV) outbreaks better than gray wolves. The K locus is located in the region that also encodes for a protein that helps defend against lung infections in mammals, thus enhancing survivability of black wolves exposed to CDV. Additionally, in areas of CDV outbreaks, wolves preferentially select a mate of the opposite color: a black wolf prefers a gray wolf for a mate, and vice versa, which allows transmittal of the black color gene to their offspring and thus a higher survival rate of their pups. Interestingly, gray wolves have higher reproductive success, so gray wolves survive better when CDV is rare or absent. It's amazing what we have learned through improved genetic analysis and wildlife disease work.

The relationship between wolves and humans is tens of thousands of years old. How is it that wolves, ancestors of the dogs that live in our homes as family members and sleep in our beds as man's best friend, have become some people's worst enemy? Wolfgang

Schleidt and Michael Shalter, in their 2003 article "Co-Evolution of Humans and Canids," hypothesized that the early human-wolf relationship was beneficial to both species. The relationship was not forged out of kindness or altruism; rather, everybody was simply trying to survive a brutish world.

These researchers postulate that sometime during the last ice age, 35,000 to 15,000 years ago, early humans may have watched a wolf pack chase and sort through woolly mammoths, musk ox, moose, or reindeer that were all much bigger than the wolves. Through good teamwork, the wolf pack chased animals to detect any weaknesses and then cut a targeted animal out of the herd. The wolves wore down their prey through long chases until the animal faltered or stopped. Wolves then began the dangerous job of killing their intended meal with their teeth. Every time a wolf kills something to eat, the wolf must grab the prey animal with its mouth and risk being stomped, trampled, kicked in the head, or gored by horns. A crippled wolf is usually a dead wolf.

Paleolithic humans didn't have the long-distance running ability of wolves, but they did have more brain power and were equipped with spears, atlatls, and bows and arrows that allowed them to kill prey from a distance without getting close and risking self-injury. They also lived in family groups, akin to wolf packs, and worked in teams. According to Schleidt and Shalter, prehistoric humans came in and finished off the wolves' cornered prey before any wolves or humans were injured. Thus, our ancestors may have assisted wolves in hunting and vice versa; this teamwork used the strength of both predators. Human hunters then drove the wolves off the kill, took what they wanted back to the cave, and left the scraps for the wolf pack. Wolves and humans shared the bounty, with each species doing their job with less risk of injury and an increased efficiency, resulting in a mutually beneficial relationship. It is an interesting twist on the confrontational wolf-human relationship as we know it today.

The timing and location of dog domestication is still uncertain. Some experts believe it occurred between 35,000 and 15,000 years ago, and that wolves and dogs may have shared a common proto-wolf ancestor. But it's complicated. What is fairly certain is that the wolf-human relationship began to deteriorate about 10,000 years ago when people additionally began to domesticate livestock and plants—and started to live a more pastoral and agrarian lifestyle. When humans became sedentary and dependent upon their live-stock and crops for food, wolves became rivals when they killed precious sheep or cattle. Wolves became hated competitors instead of collaborators, and people began telling terrifying stories of monstrous wolves. Meanwhile, the wolves that people had already domesticated evolved into Labrador retrievers and cocker spaniels, lying beside us in our beds and guarding our children and livestock.

A FEW YEARS AGO, I bumped into an older North Fork neighbor as we were both walking on a rough Forest Service road. As she began telling me a story about encountering wolves, she began to tremble, and her eyes became fearful and filled with tears. She told me that she had been out walking her friend's border collie when the Kintla Pack came trotting down the road. The dog ran to her, and she scooped up the thirty-five-pound collie in her arms and held her tightly against her chest as the seven wolves circled around them. The wolves eventually wandered away, leaving the shaking woman and dog standing in the road. She was terrified as she relived the scenario for me and lashed out, "They could have killed me!" I felt awful for her as I listened to her story unwind and saw her intense fright. But I was thinking to myself, "You know, if they wanted to kill you, you wouldn't be here to tell me the story."

My neighbor is not alone in her reaction to what she perceived as a highly threatening situation. In 2018, a young salmon habi-tat researcher in Washington State unintentionally walked into a wolf rendezvous site in mid-July, and an adult wolf began barking and howling at her. The woman was terrified, climbed a tree, and radioed her boss that she was treed by wolves. State and federal agencies went into rescue mode overdrive and sent a Department of Natural Resources helicopter into this remote area to save her. The dramatized news went viral. Headlines shouted, "Helicopter Rescues Woman Treed by a Pack of Wolves."

Typically, wolves will bark or howl when alarmed to scare intruders away from their pups or from a kill. When they do this, they sound like a cross between a barking rottweiler and a baying bloodhound. I have experienced this eerie vocalization many times, and it always makes the hair on the back of my neck stand up. But I don't worry that the barking wolves will kill me, because they could already have done so in deadly silence. They are simply telling me in clear wolf language to please leave. In the Washington wolf event, no one was hurt, and the wolves were probably wondering why, despite their clear request for the woman to leave their rendezvous site and pups, the woman instead brought in a strange whirlybird machine and armed troops to collect her. If the woman had kept on walking out of the rendezvous site instead of climbing a tree, the guardian wolf would have gone back to the pups and the incident would have been, well, without incident.

AND SO, HUMANS continue to have their own landscape of fear when it comes to wolves. From the Brothers Grimm's gruesome fairy tales to modern legislative decrees, the deep-seated fear of wolves has not been eradicated by science and factual findings. As a biologist, I am a believer in science, and I just don't understand this terror that escalates into something beyond hatred. In some people, it is a fear that runs so deep that it is impossible to deal with rationally.

Every year, many people are injured or killed by mountain lions, bears, moose, coyotes, bison, deer, and many other wild animals. By comparison, wolf attacks are extremely rare. And yet the general public doesn't hate these other human-injuring species like they do wolves. A wolf pack that brings down elk, moose, and bison would find a human to be pathetically vulnerable and easy to kill.

A couple of questions have always fascinated me: Why don't wolves kill people more often? And even more importantly, why do some people hate wolves so bitterly? If only I could figure this out. I want to put the wolves in the biological realm of mountain lions, bears, and moose, and give all of them the respect they deserve as wild animals. And yet fear and hatred of wolves is alive and well, refusing to be crushed with logic and accurate science. Just read the anti-wolf bumper stickers, attend rod and gun club meetings, rub elbows with ranchers, or read op-eds. Nowhere was this attitude more clearly demonstrated to me than when I walked into the ugliest public meeting of my career in Trout Creek, Montana.

15

······

TROUT CREEK

I T WAS NOVEMBER 28, 2018, and I was embarking on one of my
most memorable public events as the wolf and carnivore
specialist for the state of Montana. The event had been organized by Glenn Schenavar, an out-of-the-area contractor who builds
nursing homes, who was spearheading a campaign to reduce the
number of wolves in the state. We knew we were in for trouble
when our Montana Fish, Wildlife & Parks crews pulled up to the
Lakeside Resort conference center in tiny Trout Creek and couldn't
find a parking spot. Every square foot of asphalt for a quarter
mile was packed with pickups sporting bumper stickers that read
"Wolves, smoke a pack a day" and "Save 100 elk, kill a wolf." Where
did these 250 people come from? We parked our trucks in well-lit
areas to minimize the chances of our tires getting slashed.

Trout Creek is a fading timber town on Highway 200 between
Missoula, Montana, and Sandpoint, Idaho. It boasts two gas stations,
a motel, a post office, three churches, and three bars to serve its
150 residents. Losses in natural-resource jobs have been partly back-
filled in recent years with a tourism economy, thanks to the fact
that the town sits on the shore of a major reservoir on the Clark
Fork River. Trout Creekers are especially proud of their motor-
sport events, along with hunting and fishing opportunities. It's a
hopping place on summer weekends, when tourists and locals min-
gle at the Naughty Pine Saloon. In 1981, Trout Creek was named
"Huckleberry Capital of Montana" by the Montana legislature, and

the distinction is celebrated every summer with the Huckleberry Festival. People also pack into town for the snowmobile poker run and hog roast in February, and the annual ATV rally in June.

NORMALLY, TROUT CREEK is quiet in the offseason, but on this November night an angry mob had gathered to vent about wolves. The raucous noise and menacing smell of angry, drunk men hit me as I walked into the crowded conference room. The state biologists, and my boss Neil Anderson in particular, had been invited to answer questions from the public about wolves. Neil had wanted to make a short presentation, but the organizers didn't allow that. It was clear this would not be a discussion; mostly, we were there to be a target for their outrage, and maybe to answer a few questions. Neil sat at the table at the front along with State Representative Bob Brown and leaders of the anti-wolf groups the Foundation for Wildlife Management, Montana Sportsmen for Fish and Wildlife, and the Montana Trappers Association. Neil was the only person at that table who would not, given the opportunity, deliberately swerve to run over a wolf standing on the side of a road.

The large room was packed long before we arrived, and when we walked in a round of boos went up from the crowd. My faint hope that this was to be an evening of informative exchanges went up in smoke. We were all in uniform, with the traditional Fish, Wildlife & Parks (FWP) grizzly bear arm patch, and thus easily identifiable. "Wearing the bear" is a motto of departmental pride, but tonight it marked us as targets. Folding chairs were filled wall-to-wall with pissed-off people, and occupied chairs overflowed out into the hall. It was a classic small-town Montana motel conference room, with an open bar in the back that fed the free flow of conversation, a cement floor that screamed every time a chair was repositioned, and a knotty pine ceiling that amplified every decibel. The obligatory taxidermied animal trophies stared down from every wall and surrounded a large American flag. Throughout the room, people were openly wearing holstered guns, and no doubt many more concealed weapons were worn under coats.

LOOKING AROUND, I saw that there were only two empty chairs in the entire room, located about a third of the way back and against the left-side wall. I walked toward the pair of empties and asked a woman with a large drink in her hand if the seats were already taken. She looked at me through bleary eyes, smirked, and said, "Yes." I looked her in the eye, said, "Thank you," sat in the seat next to her, and waved my colleague Bruce Sterling over to take the other chair. Bruce and I weren't the bull's-eye—Neil was—but we were on hand to answer any specific questions that were within our areas of expertise: me as the wolf specialist, and Bruce as the local area biologist for the past thirty-five years. Bruce was the person responsible for the management and monitoring of all wildlife species in the area, professionally responding to a wide variety of public calls every day, and nearly everybody knew him.

In addition, there were armed FWP wardens and a FWP commissioner who had driven all the way from Missoula to participate in the meeting. As I looked around the room at the camo-clad, sidearm-packing public, I realized we were seriously outgunned. It was the only public meeting I have attended where I thought it was possible that someone could get shot. Probably someone wearing the bear. My heart rate and blood pressure soared as I took it all in.

Schenavar opened with, "The introduction of wolves has changed the traditions and cultures of Montana," which generated a round of yeahs and knowing nods. He introduced the seated panel members and their affiliations, and Neil's introduction received another round of boos. I felt sorry for Neil up there, but he was used to dealing with the public on controversial issues. Still, this was wolves. In Trout Creek. As the evening wore on and the public flogging grew in intensity, the more internally irate I became. The audience rantings got louder, people were shouting and waving their fists, and some folks got up and stormed out. That was in the first half hour. It was getting rowdy but not loud enough to cover the clatter of beer bottles being knocked over onto the concrete floor and rolling under the chairs. It reminded me of sitting in a movie theater as a kid and hearing somebody's jawbreakers spill out of the box and roll all the way down the floor to the front row.

The moderator occasionally reminded audience members to show respect for the panel members, meaning Neil, but didn't usually rein them in until after things had gotten out of hand. "Our mission is to promote ungulate recovery in areas impacted by wolf introduction," declared one panel member, followed by applause. As if this was about science and not about politics. I was thinking to myself, "First of all, these wolves were not reintroduced, they walked down here from Canada; and second, ungulate populations hadn't been devastated by wolves in northwestern Montana, where these people lived, worked, and hunted."

Bruce stood up and reported that elk numbers had been stable for the last thirty-five years since he began monitoring them. He was talking science and long experience. He added that the area was not desirable elk habitat and the winter range here wasn't what elk really needed for survival. An audience member called Bruce a liar, saying there were no elk out in the mountains now, and that years ago there had been elk everywhere, and the lack of elk was because of the damn wolves. Bruce's hackles went up at being called a liar and he responded in heated rhetoric. Game on.

A fellow yelled that all the elk had moved to private lands in the valley, where hunters couldn't get access to kill them, because the wolves were in the mountains pushing the elk out. Neil tried to explain several things that contribute to elk movements: the number of hunters had gone up significantly, elk were fleeing the onslaught of hunters' pickups on logging roads, bulls had been overharvested for several years, and logging and fires had changed the habitat over the decades. Nobody acknowledged Neil's science-based comments.

Members of the audience were hollering about reducing the wolf population down to the 150 wolves statewide that FWP had set as the limit in their federally mandated wolf management plan. Neil tried to explain that 150 wolves was the minimum number that the state of Montana had to maintain as required by the Endangered Species Act. We had to keep the number of wolves above 150 or else the federal government would take management authority away from the state of Montana and return wolves to endangered status again.

"Where the hell did that 150 number come from anyway when we've got 850 of them in Montana?!" shouted a bearded guy in a plaid shirt.

"Diane was involved in that process many years ago, so perhaps she can answer that question," Neil said as he looked at me.

My turn. I stood up from my chair on the sidelines. "That is a very good question, and I will answer it. But before I do, I have to say that when you invite us here for a discussion, and you boo us, you shout over us, and you walk out, it is not helpful to a productive conversation." I said it slowly and calmly as I stared out into the audience, daring them to meet my gaze. My comments hung in the air, and all went quiet. Then I answered the original question, going back in time to when wolves were just trickling down from Canada. I told the crowd that the wolves in northwestern Montana walked down here from Canada on their own, before the reintroductions. I knew because I was there.

In planning for the reintroductions into Yellowstone and Idaho, the federal government was required by the Endangered Species Act to determine how many wolves were needed for a viable population. They sent out surveys to wolf biologists all over North America, including me, asking that question. The consensus was that 100 wolves were each needed in Montana, Idaho, and Wyoming, for a total Rocky Mountain population of 300. But since Wyoming wouldn't play ball with federal management, the U.S.

Fish and Wildlife Service didn't delist them in Wyoming and left it up to Montana and Idaho to sort it out. So Montana and Idaho each agreed to maintain a minimum of 150, for a total of 300 between our two states, and that met with federal standards. Then I sat down. It was still quiet. Maybe they had a hard time with the math.

Justin Webb, executive director of the Foundation for Wildlife Management (F4WM), spoke up from the front table to get the meeting back onto the topic wolf reduction. He explained how people in Montana needed to trap a lot more wolves to get rid of them. He avowed that trapping, though expensive (he claimed to have spent $1,600 in fuel alone for each of the sixteen wolves he had trapped the previous year in Idaho) was the most successful method to rid an area of wolves. Montana's trapping regulations prohibited bounty payments and were the main reason F4WM had not been successful in the state. Wolf trapping had flourished in Idaho because in that state the payment of $1,000 per wolf wasn't considered a bounty, but rather a "reimbursement of expenses." The monetary support was coming from F4WM, the Rocky Mountain Elk Foundation, and the Idaho Fish and Game department. Webb was campaigning hard to get a similar program going in Montana.

Now that the discussion had turned from biology to killing wolves, the audience was pumped. I could practically see them running out tomorrow to set wolf traps everywhere, calling wolves in with varmint calls, and shooting them with glee. The audience conversation bubbled over, mostly with the typical anti-wolf and anti-government rhetoric I had heard before, but it also brought up some new tangents.

"Hell, we didn't want you to put them here in the first place, and now we're stuck trying to get rid of the bastards!"

"We are out there all the time and have lots of numbers for our deer, elk, and wolves. And our numbers don't match your numbers, and ours are right. We live here, you don't, and you don't have any idea what's going on."

"Bill Clinton is a pedophile."

"I'm afraid to let my grandkids wait at the corner bus stop because the wolves will get them."

"It all started with President Roosevelt and taxes."

"Wolf season should be 365 days a year with no bag limit."

"These wolves aren't native. These are Canadian super-wolves that weigh 175 pounds and are more aggressive."

"You're supposed to be managing these wolves for us, so why aren't you doing so?"

"The wolves are killing all the deer and elk. We can't feed our families because the wolves have eaten everything."

"How come we can't trap in the summertime like the Fish and Game biologists can?"

This last question came from a man seated about fifteen feet away from me. Neil spoke up and said, "Diane can answer that because she's probably trapped more wolves than anybody in this room." Heads swiveled toward me. I stood up for the second time that night and began to explain how important it is to avoid recreational wolf trapping during the months that federally protected grizzly bears are active—so they aren't accidentally caught—and how I set my traps in summertime with lures that are attractive to wolves but not bears in order to—

In the middle of my answer, the buddy of the guy who had asked the question began chuckling, guffawing, and shaking his head. I stopped in mid-sentence and looked at the heckler.

"What are you laughing at?" I said. "I'm answering your friend's question, and we need a little quiet here so I can finish." He stopped smirking and shut up. It got quiet again. I continued my answer until I had finished my thoughts. Then I sat down.

I've been in this wolf business a long time and I rarely uncork, but after watching Neil, Bruce, and all of us at the FWP receive so much abuse, I just couldn't help but draw the line. Somebody needed to tell them to stop acting so childishly. I don't know why it worked but it seemed to, at least for a minute, before they gathered steam and the verbal abuse began again, mostly toward my colleagues. Maybe because Neil and Bruce are big men, they were perceived as more of a threat. It was much more fun to beat up on them than on a wiry, older woman—perhaps it would be more like disrespecting Mom or something.

I didn't have to answer any more questions during the remaining hour of emotional, over-the-top rhetoric. While the discussion was unfolding, the game wardens quietly slipped out of the room and began walking along the rows of pickups, taking down license plate numbers just in case things went a bit western. They saw license plates from Priest Lake, Libby, Eureka, and other rural Montana and Idaho communities that were a hundred plus miles away.

After two hours of verbal fire directed at the FWP, the moderator wrapped up the session and thanked everybody for coming. He invited folks to stick around and chat informally if they had more questions, like an old-fashioned church potluck. Most of the crowd left, but several people stayed to talk one-on-one with Neil, Bruce, me, and the other panel members. Some were even softly apologetic of the crowd's behavior and thanked us for coming. That was the best we could hope for.

THE RHETORIC I HEARD in Trout Creek felt like something out of the Dark Ages. It was disturbing to realize that some things don't change, no matter how much effort you put in. When there are firmly entrenched prejudices perpetuated in isolated communities, it's an uphill battle to get anyone to see things differently, and after so many years of work, it's supremely frustrating to see that you're getting nowhere. And yet, in the larger picture, wolves are winning their recovery battle.

Wolves are winning because of their intelligence and resilience, and because of increasing human tolerance in an ecological landscape vastly larger than Trout Creek. Wolf populations have expanded their numbers and range in Montana, the entire U.S., and globally. Long ago, when the first wolves walked south into the

North Fork from Canada, every single wolf was critical to the survival of that precarious first pack. The death of the breeding female could mean the collapse of the budding population. The survival of every individual was crucial, and I felt the death of every single one. I never imagined that forty years after I started work on the North Fork, wolves would have resoundingly recovered and that the Montana public would legally kill nearly three hundred wolves annually for recreation, apparently without decimating the population of the thousand wolves residing in this state.

16

······

DARK TIMES
RETURN

"**A** SLAUGHTER OF WOLVES like this hasn't been seen in a century," wrote Thomas McNamee in his January 17, 2022, op-ed in the *New York Times*, talking about Montana's new laws on killing wolves.

Kill all the wolves that you can, by any means possible. Neck snares, leghold traps. Shoot them in the dark, kill them year-round, hire professional wolfers to kill them, dig pups out of dens and shoot them, and get paid a bounty for every wolf you can kill. Is this the late 1800s? No, it is Montana and Idaho in 2023, thanks to the laws passed by the conservative legislators and governors in 2020 through 2022. The old western paradigm that "the only good wolf is a dead wolf" has been resurrected, aided by the modern technology of night-vision goggles, spotlights, electronic calls, and snowmobiles and ATVs—used for running wolves down.

The Foundation for Wildlife Management (F4WM), the foundation that had its executive director at the head table at the meeting in Trout Creek, is a nonprofit that was started in Idaho in 2011, with its primary goal being to kill wolves and provide bounties (aka compensation) to wolf hunters and trappers. This group is well organized and well funded, with hungry followers and a charismatic leadership whose rhetoric incites people to go out and kill wolves.

f4wm has partnered with Idaho Fish and Game to help pay "compensation of expenses" to successful wolf hunters ranging from $500 to $2,000 per wolf. It seems unsavory that a state wildlife management agency would team up with a strongly anti-wolf NGO to pay for dead wolves when the public supported reintroducing wolves, at great expense, twenty-five years previously. The rallying cry of "no wolves here" is growing louder.

The Idaho legislature barred the state Fish and Game department from getting involved with wolf reintroduction back in 1995, and the wolves ended up being reintroduced in that state by the Nez Perce Tribe; tribal elders blessed crated wolves on the tarmac in Missoula after the captured wolves had been flown down from Canada for the reintroductions. However, despite the state's political posturing, the wolf writing was on the wall even back then, and the Idaho legislature created a seven-member Wolf Oversight Committee that helped develop a wolf management plan prior to the 1995 reintroduction in Idaho's central wilderness.

Seven years later, in 2002, the Idaho legislature adopted the Idaho Wolf Conservation and Management Plan, the federal U.S. Fish and Wildlife Service accepted it, and Idaho tiptoed into the era of wolf management. In 2005, a collaboration between the Nez Perce Tribe and Idaho Fish and Game reduced the geographic area the tribe would manage. The formal partnership between the tribe and the state ended in 2016, after the reintroduction programs in Idaho, Wyoming, and Montana had been deemed successful and wolves in those states dropped off the endangered species list. Since the state of Idaho has taken over wolf management, a definite anti-wolf sentiment has pervaded the state's political perspective.

In 2020, Idaho Senate Bill 1247 established "wolf-free zones" by expanding wolf hunting opportunities year-round in many hunting units. The following year, Idaho Senate Bill 1211 (aka the Wolf Management Bill) went into effect, which allows for killing 90 percent of Idaho's wolf population. This goal, to decrease Idaho's wolf population from 1,500 wolves to 150 wolves, will be executed through contracted professional wolf hunters and allow a year-round hunting season with no bag limits. The bill also allows hunters to use dogs to pursue wolves. With wolf management now in the state's hands, the legislature increased wolf control funds from $110,000 to $300,000, paid through the Idaho Fish and Game department, which hires private contractors to kill wolves.

In addition, the state legislature earmarked $200,000 for the Idaho Fish and Game department's contribution to wolf bounties, which are jointly paid for with the F4WM and the Rocky Mountain Elk Foundation: $2,500 per wolf where wolves are chronically preying on livestock, $2,000 per wolf where the state Fish and Game department thinks that elk are below management objectives, $1,000 per wolf in northern Idaho, and $500 in the rest of the state. Twenty-five years after wolves were reintroduced to Idaho as an endangered species, they were once again in the firing line.

MEANWHILE, IN MONTANA, during the six-month wolf harvest season (September 4 to March 15), trappers can use neck snares to slowly strangle wolves and leghold traps that keep wolves alive until the trapper arrives to kill them. Hunters can lure wolves with baits and electronic calls and hunt wolves at night with artificial lights, thermal imaging, and night-vision scopes. Each licensed person can kill up to twenty wolves (ten by hunting and ten by trapping). And there is no limit to the number of wolf licenses that can be sold. Wolves may be hunted and killed with archery bows and arrows, crossbows, rifles, handguns, muzzleloaders, and shotguns loaded with buckshot or slugs.

Hunters must notch the wolf tag with the date of kill and report it to Fish, Wildlife & Parks. Most hunters attach the notched tag to the wolf and pack the dead wolf out, but if a hunter doesn't want to touch the wolf, it is legal for the hunter to notch the tag and simply drop it onto the dead wolf without touching the carcass.

A successful hunter is not required to remove any part of the dead wolf from the kill site and can leave the carcass rotting in the field. In any other big-game species, this would be wanton waste of an animal and subject to legal penalties. This state-sanctioned disrespectful treatment of harvested wolves gives the message that wolves are vermin and not a valued species in the eyes of sportsmen and sportswomen.

Several new antipredator laws were passed by the Montana legislature between 2020 and 2022, mostly by legislators from the Trout Creek area. Montana Senate Bill 267 (aka the Wolf Bounty Bill) authorized paying reimbursement costs incurred related to killing wolves through trapping and hunting.

The Montana legislature is returning to darker times in terms of wolf management policies. Between 1870 and 1877, approximately 34,000 wolves were killed in Montana and southern Alberta. The legislature of the Montana Territory began paying bounties for wolves, coyotes, and other predators in 1883. A full wolfskin brought a bounty payment of $1. By the end of 1884, bounties were paid on 5,540 wolves, 1,774 coyotes, 586 bears (species not specified), and 146 mountain lions. Many, if not most, of the animals were killed by strychnine—advertised in the September 3, 1901, *Billings Gazette* as "The Only Poison Made That Never Fails to Kill." Elmer E. Crawford commented in this article, "We have gotten as high as twenty-four grey wolves with one ounce of the strychnine, and the best feature of it is they are found close to the bait." In 1905, the bounty rates increased to $10 for an adult wolf and $3 for a pup. By the late 1920s, after forty years of intense government-funded wolf killing, there were few wolves left in the Lower 48, with the exception of Minnesota.

Montana dropped its wolf bounty payment in 1933 because the extirpation job had been successful. Bounties were paid on 80,000 wolves in Montana from 1883 to 1918 and 30,000 wolves in Wyoming from 1895 to 1917. Let us hope that the Montana legislature doesn't resurrect the Montana wolf mange law of 1905: "An Act to provide for the extermination of wolves and coyotes by inoculating the same with mange, and to place such duties under the charge of the State Veterinarian." Wolves were trapped, infected with mange, and then released in the hopes they would spread it to the rest of their packmates.

In 2021, Montana Senate Bill 314 increased wolf harvest, with the intent to reduce Montana wolf population to just fifteen breeding pairs, or approximately 150 wolves. This bill allowed unlimited killing of wolves by a single license holder in some areas of Montana. It also allowed wolf night hunting with artificial lights, night-vision scopes, and use of bait, none of which had been previously allowed in the past century.

Montana House Bill 224 allowed the snaring of wolves as a method of take, also jeopardizing many nontarget species and dogs. I saw a comment by an anti-wolfer online supporting this bill that said, "It is finally time to move on from 'Smoke a pack a day' to 'Choke a pack a day.' I wish everyone luck." I know of a bird hunting couple whose two treasured bird dogs strangled to death in the owners' presence as they unsuccessfully struggled to free their dogs, which had been caught in wolf neck snares. They were legally bird hunting on Montana state lands, and a trapper had legally placed his snares on public lands.

Montana House Bill 225 lengthened wolf trapping season by one month. This extension is problematic for grizzly bears as well, because grizzly bear researchers have documented that due to climate change, grizzlies are going into their dens later and emerging sooner. This increases the chances of catching an adult grizzly, or worse yet a cub, endangering the bears and the trapper.

An incident occurred in the Swan Valley of northwestern Montana in December 2018 where a grizzly cub was caught in a trap, with the angry mother bear and the other cub running freely around the trap site. The trapper heard the cub bawling as he walked in to check his trap. He recognized the sound and backed away before the mother bear detected him, and he called Montana FWP.

Releasing the cub unharmed, on-site, involved numerous FWP armed staff, bear specialists, and a helicopter hazing the mother grizzly away so the bear crew could safely release the cub. It was a very tense situation for all involved. If the trapper hadn't paused on his walk in to check his trap, he could easily have been severely or fatally mauled by the enraged mother grizzly. We are likely to see more such incidents with more unrestrictive wolf trapping regulations.

The Montana state wildlife commission established harvest thresholds which totaled 450 wolves statewide, approximately half of the wolf population. If the 450-wolf kill is reached, that triggers a commission review but does not necessarily close the season. "A harvest of 450 wolves shall initiate a commission review with potential for rapid in-season adjustments to hunting and trapping regulations. Thereafter, the commission shall be similarly re-engaged at intervals of additional 50 wolves harvested."

In 2020, hunters and trappers killed 329 wolves in Montana. But despite the expansion of wolf killing laws, Montana's state wolf harvest decreased for the second year in a row, with 273 wolves killed in 2021 and 258 killed in 2022. We can presume that the amount of effort to kill wolves was likely similar (or even higher, due to longer seasons and more methods to kill) in the past three years. Logic suggests there were, therefore, fewer wolves out there to kill. Someone asked me if perhaps the wolves were getting smarter because of increased hunting and trapping pressure. My thoughts are that the wolf who learns the most is dead, a one-time learning opportunity. If packmates see one of their own trapped or shot, do they learn from it? I really don't know.

During the open comment period, the state received 26,000 public comments on the proposed aggressive wolf kill legislative changes— by far the largest number of comments ever received on any issue.

Approximately 80 percent of the comments were in opposition to increasing wolf hunting and trapping. The state of Montana was sued by several conservation groups to reverse the anti-wolf bills. Conservation groups petitioned the U.S. Fish and Wildlife Service to consider relisting wolves under the Endangered Species Act to afford protection. The pro-wolf organizations lobbied at the state legislatures, put up pro-wolf billboards, and wrote op-eds. Wolf supporters came out in Montana, Idaho, and nationally in support of science-based conservation, not politically charged wolf kill agendas.

Following a February 10, 2022, court order, gray wolves in the contiguous 48 states and Mexico—except for the Northern Rocky Mountain population—were once again protected under the Endangered Species Act as threatened in Minnesota and endangered in the remaining states. Wolves in the Northern Rocky Mountain area were not federally listed and remain under state management. Once again, the status of wolves has been bandied back and forth federally and by the states at the whim of political agendas, excluding scientific parameters.

BUSINESS OWNERS IN Gardiner, Montana, and other gateway communities to Yellowstone National Park depend on tourism and wolf watchers to support their businesses. They do not want to see more wolves killed and their livelihoods threatened. In 2022, twenty-five of Yellowstone's one hundred wolves were killed by hunters and trappers. Several of these dead wolves had been radio-collared and were important to the ongoing wolf research efforts. The wolf killings created a large public outcry from people across the U.S. Some of the killed wolves were well-known favorites to photographers and wolf watchers. What goes without any public acknowledgment is that approximately one-third of the entire wolf population is legally killed by humans every year in Montana and Idaho. That's three hundred wolves every year in Montana, and approximately five hundred wolves annually in Idaho. Although these animals are not superstars like the Yellowstone celebrity wolves, they mate, kill deer, produce pups, play, starve, cooperate, howl, and do all the same things as the Yellowstone wolves; it's just that nobody sees them—except through rifle crosshairs or in traps.

Wolves in Wyoming were removed from Endangered Species Act protections in 2017. There is a managed annual hunting season in the Wolf Trophy Game Management Area, which covers those parts of Wyoming where wolves are classified as trophy game animals. In 85 percent of the rest of the state, wolves are considered predatory animals and can be killed without a license at any time, year-round, with no limit on the number of wolves that can be killed. Wyoming's goal in 2021 was to stabilize the wolf population at roughly 160 wolves in the Wolf Trophy Game Management Area, remaining above the federal minimum recovery criteria.

THE STATE OF WISCONSIN opened their 2021 wolf hunting and trapping season in February 2021. The state set the harvest quota at 119 wolves out of the state's roughly 1,000 wolves. But 218 wolves were reported killed in just three days, and the season was shut down as quickly as Wisconsin state officials could get it closed. Nearly twice as many wolves were killed in three days as was the allowed limit. Eighty-six percent of the wolves killed were taken with the aid of hunters' dogs. Hunting wolves with dogs is particularly effective; it seems a cruel twist to use the domesticated cousin of the wolves to run them down and kill them.

IN SEPTEMBER 2022, the U.S. Fish and Wildlife Service announced that it was beginning a twelve-month review to determine if "potential increases in human-caused mortality may pose a threat" to gray wolves. "The Service also finds that new regulatory mechanisms in Idaho and Montana may be inadequate to address this threat. Therefore, the Service finds that gray wolves in the western U.S. may warrant listing."

Under the Endangered Species Act, a species must be listed if it is threatened or endangered because of any of the following five factors:

- present or threatened destruction, modification, or curtailment of its habitat or range;
- overutilization [of the species] for commercial, recreational, scientific, or educational purposes;
- disease or predation;
- inadequacy of existing regulatory mechanisms; or
- other natural or manmade factors affecting its continued existence.

As a wildlife professional, it has been drummed into my head to manage wildlife at the population level, and to disregard the importance of individual animals. But not all wolves are created equal. An older wolf or a breeding animal contributes appreciably more to the stability, experience, and longevity of a pack, guiding them in making good choices. Older wolves are knowledgeable leaders and the glue that holds the pack together.

The vast majority of wildlife studies and management protocols focus on growth rate (called lambda), population size, home range size, birth rate, and overall mortality, etc., and disregard impacts on smaller scales such as the impact of removing the leaders of family units in a social species like wolves. When a breeding male or female is killed by humans, it is not merely one more wolf killed. Kira Cassidy, of Yellowstone Wolf Project, and her colleagues published their findings in January 2023, suggesting that which wolf is killed by humans can have a significant negative impact on pack persistence and reproduction. Breeding wolves are the most important members to keep a pack functioning, and old wolves (wolves six or more years old) also disproportionately contribute to pack survival through their years of life experience—what we would call wisdom in humans. "You kill the wrong wolf at the wrong time, that pack could blink out or won't reproduce," said

Doug Smith, recently retired lead wolf biologist with the Yellowstone Wolf Project. As long as levels of human-caused mortality are conservative, wolf packs usually recover eventually, and those that disappear are replaced by other wolves, so the overall population is likely to remain stable. Problems arise when humans kill so many wolves that natural replacement cannot keep up.

Cassidy has also published results on the value of specific age classes and how older wolves may contribute significantly more than younger animals to the social structure of a pack. The Yellowstone team also documented that larger packs are better able to survive the loss of a breeder killed by humans. Part of the problem with human-caused wolf mortalities, such as hunting and trapping, is that they often occur during the breeding season or when females are pregnant—or are more likely to occur in clusters where hunters kill more than one member of a pack—whereas natural mortalities are more evenly spread within packs and throughout the year.

And so, there are social consequences to wolf killing that are not assessed in standard population studies of wolves, which do not consider the contributions made by individual wolves, the disruption of the family group when a breeder is removed, and social stability within a pack when elders are no longer there to impart their sagacity to the younger wolves. Which leads one to question whether wolf hunting and trapping can be acceptable or ethical, except in very specific circumstances. I know too much about the intelligence, social dependence, and cooperative nature of these superb carnivores to be able to fathom why their lives should be extinguished for sport. I agree with Barry Lopez's musings in his book *Of Wolves and Men*, where he writes that wolves are often made scapegoats for the qualities we most despise and fear in ourselves.

I HAVE NOTICED the increasing publication of good wolf research in popular literature, as well as in scientific journals, resulting in more people reading about wolf science in a more understandable format. And some good has come out of all the recent anti-wolf rhetoric: it has united wolf supporters across the country to speak up, join in legislative and congressional battles, and become more politically active to promote wolf conservation through science-based decision making.

It seems that the citizens of Montana were tolerant of more reasonable wolf harvest laws a few years ago, even if they didn't like hunting and trapping. But the irrational and extreme wolf kill laws from 2020 to 2022 launched the formerly quiet majority into action. By 2022, nonprofit legal and conservation organizations were litigating through the courts to try to legally challenge and overturn arbitrary laws enacted in the previous three years. Throughout all this turmoil, relisting, and delisting, the wolves have persisted.

So where does this leave wolves, other than in the literal and figurative crosshairs of the public and policymakers? I cannot say, but I can hope that the wolf's future survival is guaranteed, not necessarily by human endeavors, but by the wolf's impressive resiliency.

17

.....

RESILIENCE

ESILIENCE IS KEY to the survival of the wolf, *Canis lupus*. Wolves are truly remarkable animals ecologically, morphologically, behaviorally, and socially. They can survive in any habitat, in any climate, and on any diet. Their body size can evolve over generations to become larger or smaller based on the size of the prey they kill. Bison- and moose-killing wolves are bigger, whereas wolves that prey mainly on white-tailed deer and beavers are smaller. A breeding female produces a litter of six pups every year, on average, which can quickly make up for wolves that disperse or die.

Wolves were first protected in the 1970s in North America, and subsequently that protection has expanded worldwide, allowing wolves to repopulate places where they haven't been seen in two hundred years. Of all mammals, wolves are second only to humans in their historic global distribution. Currently, wolves live in much of Canada, Alaska, parts of the lower 48 states, Mexico, most of western Europe (even in the Netherlands, Denmark, and Belgium), Israel, and India. We humans have done our best to trap, poison, and shoot them off the planet; in nearly all studied wolf populations that encounter humans, people are the most significant cause of wolf deaths. Yet they persist and flourish if simply given a little protection, and they can repopulate an area remarkably quickly if they are not the target of a massive reduction campaign. The repopulation of my North Fork, much of the western U.S., the

Great Lakes states, parts of Canada, and much of western Europe in just forty years is proof of wolves' amazing resilience to human efforts to eradicate them.

ON APRIL 26, 1986, the fourth reactor exploded at the Chernobyl nuclear power plant in Ukraine, creating the worst nuclear disaster in human history. Chernobyl caused four hundred times more radiation damage than the atomic bomb the U.S. dropped on Hiroshima in 1945 and was ten times more devastating than the Fukushima nuclear disaster in 2011. Dozens of people died immediately, and some estimates are in the tens of thousands for future deaths due to cancer from the radiation. Approximately 116,000 people were immediately evacuated from the Chernobyl Exclusion Zone (CEZ) within a nineteen-mile radius of the power plant, permanently evicting everyone in a one-thousand-square-mile area. But what about the wildlife? Flora and fauna also suffered in the initial aftermath of the explosion. Thousands of pine trees died instantly, and the brown tree skeletons were dubbed the Red Forest. No one knows how many wildlife species and plants died initially, because human researchers were not allowed in the CEZ to document the destruction.

However, more than thirty-five years later, Mother Nature is reestablishing a thriving ecosystem and de facto wildlife sanctuary in the absence of humans. Biodiversity of plants and wildlife has increased. Seventy percent of the area is now covered by wild forest, significantly more woodlands than before the nuclear explosion. Humans are strictly prohibited from living in the CEZ, although some scientists do visit occasionally to monitor or view the area. Trees grow through building roofs. A Ferris wheel and a pack of bumper cars sit frozen in place by rust and vegetation. Wolves stalk prey at night along streets that have become overgrown and now function as wildlife trails.

This is not to say that plants and wildlife are not affected. Some scientists report abnormal sperm from many bird species and

partial albinism among barn swallows; birth defects, DNA mutations, and strange coloration in amphibians; insect deformities; and
radioactive moss, lichens, and fungi. Other researchers claim that
wildlife is thriving in the Chernobyl area despite the high levels of
radiation. Two hundred species of birds have been spotted in the
radiation zone, including the rare and endangered greater spotted
eagle. Scientists also report that even after three decades of chronic
and continuous radiation exposure, the CEZ supports an abundant
large mammal community and these large mammal populations
within the CEZ are now similar in abundance to those in four nearby
uncontaminated nature reserves.

Trail cameras and recent surveys have shown an increase in numbers of wolves, wild boars, brown bears, lynx, beavers, raccoon dogs,
foxes, bison, moose, and roe deer. The endangered Przewalski's
horse had gone extinct in the wild, with the last wild horse seen in
the Gobi Desert in 1969. Zoos and captive facilities have kept the
species alive since then. In 1998, conservationists introduced thirty-
one Przewalski's horses into the CEZ, since it was presumably now
a refuge from humans. Intense poaching for horsemeat in 2004
through 2006 destroyed much of the herd, but protection measures were put in place, and in 2020, 150 horses were counted in a
variety of herds and social groups. Introducing a very rare, endangered species into a nuclear disaster zone is a bold conservation
approach, but the horses appear to be rebounding well.

Forty to fifty wolves call the CEZ home, and wolf abundance
there may be seven times higher than in surrounding areas occupied by people. That may be due, in large part, to wolf persecution.
It could also be that wolves are less affected by the radiation
because they have large territories and move around a lot, wandering out of contaminated areas and into radiation-free areas—or
perhaps wolves have unique resilience to whatever comes their way.
One Chernobyl researcher, James Beasley, claims, "The preliminary
density estimates that we are seeing suggest that in Chernobyl the
density of wolves is much, much higher than even Yellowstone."

A young male wolf was radio-collared in the CEZ in November 2018. He then wandered for 229 miles over a twenty-one-day period before his collar stopped transmitting. He was likely born in the CEZ after several generations of radiation exposure to his ancestors. We don't know how much radiation damage he experienced, but he successfully made a living, looked like a wolf, acted like a wolf, and dispersed to a new wolf population. He may well be out there breeding now. His story suggests that the CEZ may serve as a source population for rebuilding wildlife populations outside of the CEZ that have been diminished as a result of human hunting, trapping, and land use in the surrounding areas. This raises questions about the potential spread of radiation-caused genetic mutations to populations in uncontaminated areas. Chernobyl radiation may turn out to be the gift that keeps on giving. Time will tell.

Is Chernobyl a radioactive wasteland reeling from chronic radiation, or a post-nuclear paradise with thriving populations of animals and other life-forms? Probably a bit of both. Chernobyl was considered a biological desert immediately after the disaster but is valued now as an area to study for biodiversity and conservation. Scientists are not saying radiation is good for animals, but they are saying that, apparently, human habitation is worse.

HOW WILL WE move forward in the future, balancing the needs of humans and wildlife in the landscape? Some claim that the return of wolves is a disaster, and some say that wolf recovery has brought about restoration of ecosystems and laud the rewilding of formerly wolfless terrain. And what of my own experience with wolves, and the resiliency of the wolves that I have studied? I was going to say "of the wolves that I have known," but a human can never really get to know a wild wolf, except through glimpses into how this wily and elusive canid makes a living.

The very first wolf we studied, Kishinena, eked out a living as a loner for at least two years before she found a mate—who had also survived some kind of trauma, likely a trap, as evidenced by his missing outer toe. He died when Kishinena's pups were barely weaned, and somehow, she managed to hunt and scavenge enough meat to feed herself and seven endlessly hungry growing pups. I am still amazed by her perseverance, intelligence, and good luck.

Female wolf 8963 was missing the last third of her tail from an unknown trauma and, at sixty-two pounds, was the smallest adult female wolf I captured for the entirety of our research. She was half the weight of her studly 121-pound mate, 8705, whom I trapped along with her when I radio-collared them both in August 1989. As the only breeding female, she was the pack's most important member—and apparently the smartest or the luckiest. All nine of her packmates were illegally poisoned up on the Wigwam Flats in March 1991, leaving her to raise the pups she would soon give birth to all on her own.

Six weeks later, my pilot Dave and I looked out of the Cessna windows at eye level with 8963 as she walked along a ridgetop above McLatchie Creek, her milk-engorged teats swinging below her belly as she searched for food for her pups. She managed to raise three pups, unaided, through the fall. Her small family found a new breeding age male, and the pack grew with more pups over the next two years. Wolf 8963 was legally shot by a hunter, Bryan, in September 1993 in British Columbia. He called my number on the inside of the radio collar to let me know he had the collar and the wolf. We talked for a few minutes, and then I told him that I would like to come up and meet him and give him some information about this hardy little female.

I drove up to his modest home east of Fernie, British Columbia. Bryan invited me in, showed me the collar, and told me his story. He had been out elk hunting when this wolf came into view, trotting along a game trail. He watched her through his scope, thrilled to see the wolf. When she paused, he shot her cleanly with one shot. He told me that this was the most meaningful hunt he had ever had, his best trophy, better than all his big bull elk. He'd never had a chance to shoot a wolf before. He admired her beautiful light gray coat, and how she suddenly appeared out of the lodgepoles like a ghost, unaware that he was there. He was going to have her hide tanned and preserved as a memory of that hunt. He hoped to see more wolves around in the future and hoped that his son might have a chance to see some too—and maybe even shoot one.

Although I don't understand the desire to shoot a wolf, I was pleased that Bryan valued this animal and wanted to continue to have wolves on the landscape. He didn't shoot her out of hatred or to save the elk. I pulled out 8963's data sheets that I printed for him: her capture sheet, her age and weight, a map with the outline of her territory, the story of her mates, how many pups she produced, and her pack history. This diminutive wolf had led a big life. His story and mine confirmed our mutual respect for her.

Wolf 8756 was another remarkable wolf. I first captured her as a feisty, five-month-old female pup in Glacier National Park in October 1987. I walked up to her along the eight feet of chain stretched out from the anchor as she strained in the trap to get away from me. As I came within three feet of her and was lining up my jab stick to poke her in the rump with the tranquilizer, she whirled and faced me, and then charged me full-on. Shocked, I ran backward as fast as I could, and she chased me for sixteen feet (eight feet either side of the anchor). I tripped over a log just as she reached the end of the chain, where she was drawn up short, probably saving me from being bitten. I picked myself up, looked at that black-as-a-bear wolf—her fearless puppy gaze staring steadily at me—and thought to myself, "You're going to be unstoppable someday."

I sedated 8756, fitted her with an oversized, padded radio collar that she would grow into as an adult. Two years later, she whelped her first litter. She turned out to be the most prolific and longest-lived wolf in our study, as well as in most other research projects I've reviewed. We were able to catch 8756 and replace her radio collar four more times during her twelve years of life, as she stayed with the same pack that she was born in. During her later years, she was replaced with a younger breeder, but 8756 was allowed to remain in the pack long after her only useful contribution was probably to be a babysitter at the den and rendezvous site.

When she was last captured by Tom Meier and me on May 16, 1999, she was near the traditional den area but had no milk in her teats—and hadn't borne pups. But she was tending the new breeding female's pups. Wolf 8756's toenails were so long that some were curling over the ends of her toes, indicating that she wasn't walking around much, let alone hunting. She was just grandma, babysitting the next generation of strong youngsters. She was pure white in color, a vastly different wolf than the solid black pup she had been twelve years earlier. I took photos of her at every capture as she gradually went from coal black to salt-and-pepper, then to a lovely smoky blue-gray and to pure white her last couple of years. Six months later, she wandered ten miles away from her pack and died of old age, alone, in November 1999, a true survivor. I admired the pluck of this canine queen.

Handsome Sage, very likely one of Kishinena's sons, survived through many hunting seasons and daily confrontations with life. He found a mate, became a successful breeder, and nearly died while being held by a trap for days in wicked winter weather. We freed him, but he never regained full use of the injured foot. And yet he persisted in defending his mate, pups, and territory on three good legs.

These wolves are tough. As soon as protections were put in place to reduce wolf deaths from humans, wolf populations increased and expanded globally. Wolf proliferation succeeded by simply stopping humans from shooting, trapping, and poisoning them. With a few exceptions (Yellowstone, Idaho, Mexican wolves, red wolves), wolves repopulated areas by themselves and spread across Canada, the western U.S., the midwestern U.S., and most of western Europe.

MY EARLY FOCUS on a few intrepid wolves transformed into a forty-year journey documenting one of the most successful endangered species recovery stories ever told. Recently, however, times have changed, and toxic anti-wolf sentiments have resulted in new laws that allow wolves to be killed indiscriminately. The pendulum is swinging back to the wolf eradication days of the late 1800s. Yet the wolves persist, resilient and clever. It is up to us if the wolf will continue to exist or if its howls will be silenced. When we hate and fear them, we eradicate them. When we love and value them, we protect them—and they return. Wolves must live somewhere in between these extremes. As I narrowed my professional focus on wolf recovery, I became aware of the critical role that listening, education, and outreach played in my work—as well as in wolf recovery everywhere. Wolf management is people management. Period.

This never-ending challenge of balancing carnivore and human coexistence is the tightrope that I have had to walk my entire career, with much angst and passion. Ultimately, the wolf's fate is in our hands; it can all change again with our social caprices. My hope is for a more tolerant world, with wolves living out their lives as a valued wildlife species. We can live without wolves, but the world is a much richer place with wolves in it.

FURTHER READING

BOOKS

Lopez, Barry H. 1978. *Of Wolves and Men*. New York: Scribner.

Mech, L. David. 1970. *The Wolf: The Ecology and Behavior of an Endangered Species*. Garden City, NY: Natural History Press.

Musiani, Marco, Luigi Boitani, and Paul Paquet, eds. 2009. *A New Era for Wolves and People: Wolf Recovery, Human Attitudes and Policy*. Calgary: University of Calgary Press.

Shivik, John A. 2014. *The Predator Paradox: Ending the War With Wolves, Bears, Cougars, and Coyotes*. Boston: Beacon Press.

Smith, Douglas W., Daniel R. Stahler, and Daniel R. MacNulty, eds. 2020. *Yellowstone Wolves: Science and Discovery in the World's First National Park*. Chicago: University of Chicago Press.

Thiel, Richard P., Allison C. Thiel, and Marianne Strozewski, eds. 2013. *Wild Wolves We Have Known: Stories of Wolf Biologists' Favorite Wolves*. Ely, MN: International Wolf Center.

ARTICLES

Boyd, Diane K., David E. Ausband, H. Dean Cluff, James R. Heffelfinger, Joseph W. Hinton, Brent R. Patterson, and Adrian P. Wydeven. 2023. "Chapter 32: North American wolves." In *Wild Furbearer Management and Conservation in North America,* vol. ii, section i: Canids, edited by Tim L. Hiller, Roger D. Applegate, Robert D. Bluett, S. Nicky Frey, Eric M. Gese, and John F. Organ, 32.1–32.68. Helena, MT: Wildlife Ecology Institute, doi.org/10.59438/FYHC8935.

Cassidy, Kira A., Bridget L. Borg, Kaija J. Klauder, Mathew S. Sorum, Rebecca Thomas-Kuzilik, Sarah R. Dewey, John A. Stephenson, et al. 2023. "Human-caused mortality triggers pack instability in gray wolves." *Frontiers in Ecology and the Environment* 21, no. 8: 356–362, doi.org/10.1002/fee.2597.

Meyer, Connor J., Kira A. Cassidy, Erin E. Stahler, Ellen E. Brandell, Colby B. Anton, Daniel R. Stahler, and Douglas W. Smith. 2022. "Parasitic infection increases risk-taking in a social, intermediate host carnivore." *Communications Biology* 5, no. 1: 1180, doi.org/10.1038/s42003-022-04122-0.

Schleidt, Wolfgang M., and Michael D. Shalter. 2003. "Co-evolution of humans and canids: An alternative view of dog domestication: *Homo homini lupus?*" *Evolution and Cognition* 9, no. 1: 5–72, nldogs.com/wp-content/uploads/2012/10/coevolution03.pdf.

ACKNOWLEDGMENTS

WE LEARN THROUGH storytelling and reading. This book came about through decades of stories, research, and experiences. It has taken a village to build this memoir, and I am indebted to so many for a lifetime of compelling opportunities and encouragement.

Many mentors helped me find my path through the wolf woods and the scientific world when I was going astray. They inspired me professionally and deterred me from leading a feral life. Mostly. Thank you Dave Mech, Jane Packard, Steve Fritts, Bob Ream, Dan Pletscher, Mike Fairchild, Ursula Mattson, Ed Bangs, and Paul Paquet for introducing me to the world of wolf research and allowing me my freedom to make discoveries and mistakes. Adventure is simply catastrophe in retrospect.

The important work of the Wolf Ecology Project would not have been possible without the scores of volunteers and technicians. Over the years, I was blessed to work with many dedicated and tireless volunteers, including Kurt Aluzas, Wendy Arjo, Mike Bader, Lou Berner, Russ Beuch, Matt Black, Andrea Blakesley, Pam Broussard, Joe Butler, Ben Conard, Randall Cooner, Bill Falvey, Pat Finnegan, Blair French, Sharon Gaughan, Tom Gehring, Denny Gignoux, Sue Habeck, Amanda Hardy, Patrick Heins, Ann Henry, Mike Jimenez, Lynn Johnson, Steve Johnson, Heather Johnston, Jamie Jonkel, Steve Kloetzel, Jina Mariani, Doug McAllister, Neil Meyer, Gray Neale, Brian Peck, Andrea Peterson, David Pilliod, Jill

Reifschneider, Denise Roth, Doug Ruhman, Carol Schmidt, Jesse Sedler, Mike Sickles, Joe Smith, Heidi Svoboda, Paula White, and Rosalind Yanishevsky. I've undoubtedly forgotten someone, and I apologize for that. To this long list, I must also thank the Border Grizzly Project and all of its volunteers and technicians, and especially Canadian researchers Bruce and Celine McLellan.

The Moose City Corporation, and George Ostrom in particular, generously provided housing for our staff and volunteers. I thank Moose City itself for keeping me safe and deeply happy, and for giving me the truest home I've ever known. The half-mile-long hay meadow was a perfect runway for our Super Cub and Cessna 182 for our wolf tracking flights. Many thanks to Dave Hoerner at Red Eagle Aviation for many years and thousands of hours of safe flying, fun aerial adventures, and sky-high soaring above the Rocky Mountains. Thank you also to Strand Aviation for getting us through our first couple of years of telemetry flights, before Dave Hoerner took over that role.

Huge thanks to all North Forkers, alive and passed away, who were instrumental in tolerating those early wolves and allowing them to live undisturbed in the beautiful North Fork Valley. Thanks, also, for tolerating the wolf crew for occasionally interrupting your lives. You were friends, sometimes critics, sometimes both, helping us with roadside repairs, joyous gatherings, and telling us when and where you saw wolves.

Glacier National Park helped support our efforts with logistics and backcountry cabin use. Their staff helped with wolf collaring, babysitting tranquilized wolves, and memorable skiing and hiking adventures. Special thanks to my favorite rangers: Jerry DeSanto, Scott Emmerich, and Regi Altop.

Thank you, Dick Thiel, Allison Thiel, Marianne Strozewski, and the International Wolf Center for generously allowing me to adapt the essay I published in their book, *Wild Wolves We Have Known*, to be used in this book as my chapter entitled "Sage."

ACKNOWLEDGMENTS

I am grateful to dear friends and colleagues who helped with edits, information, and moral support, including Karen Atha, Mary Baker, Asta Bowen, Chris Brick, Mike Dever, Joe Fontaine, Connie Marmot, Francesca Marucco, Carter Neimeyer, Jessie Opel, Toni Ruth, Doug Smith, Bruce Sterling, Bridgett vonHoldt, and Paula White.

My brothers, Terry and Jeff Boyd, provided brotherly love, suggestions, and encouragement. My parents, Harold and Verna, passed away decades ago but raised me to be independent and to believe I could be anything I wanted to be.

I am so blessed to have Jane Billinghurst as my editor, and Greystone Books as my publisher. Jane, you are a consummate professional editor as well as a kind soul guiding me through this memoir writing process. It is a lot of work. I cannot thank you enough for your wisdom, and I hope that creating this book has been a fun ride for you, too.

Many thanks to Doug Chadwick for his firm nudges propelling me into writing this story. You offered sage, writerly advice when I most needed it—sometimes while I was skiing with you, Karen, and our dogs. I'm so grateful for your honest and thoughtful foreword.

And of course, none of this would be possible without the wolves. Kishinena, Sage, Phyllis, and all the wolves who tolerate our human foibles in their otherwise sane and wild world. I'm sorry for pinching your toes and eternally thankful for your insights. Long may you howl.

INDEX

Note: "plate" followed by a number refers to the photo pages.

Columbia Falls, 55, 56
Cut Bank, 87
Dutch Creek, 73, 74
Eureka, 191
Flathead Lake Biological Sta-
    tion, 17
Flathead River, 31, 137
Flathead Valley, 159
Gardiner, 199
Greenough Park, 134
Helena, 126, 128, 136
Hidden Meadow, 164
Kalispell, 32, 34, 38, 66, 133, 134,
    135, 174
Kintla Lake, viii, 3, 122–23
Kishinena Patrol Cabin, 52
Lakeside Resort conference
    center, 184
Libby, 181
Lolo, 133
Lower Sage Creek Road, 80
Marias Pass, 96
Montana Territory, 196
Naughty Pine Saloon, 184
Ninemile drainage, 87
North Fork Road, 44, 55, 92, 164
Northern Lights Saloon, Pole-
    bridge, 119
Polebridge Ranger Station, viii,
    73–74, 113
Red Lake, 20
Sage Creek, 40, 48
Seeley Lake, 136
Spotted Bear Ranger Station, 133,
    135, 138
Stanford, 30
Starvation Ridge, 123
Stevensville, 128
Sullivan Meadow, 112
Swan Valley, 197
Tepee Lake Road, 114
Trail Creek border crossing, 32,
    33, 41, 55

Trout Creek, 183, 184–85, 186, 191
Upper Sage Creek Road, 48
Whale Creek Road, 44
Montana Fish, Wildlife & Parks
    (FWP), 128, 138, 173, 184, 185, 186,
    188, 195, 198
Montana Sportsmen for Fish and
    Wildlife, 185
Montana Territory, 196
Montana Trappers Association, 185
Moose City, Montana
    author as caretaker of, 41, 43, 89
    and border relations with
        Canada, 55
    cabins at, 34, 41, 43, 89
    filming of Heaven's Gate at, 34–35
    hay meadow/runway at, 38, 101,
        116–17, 159
    headquarters of Wolf Ecology
        Project (WEP) at, 63, 111, 113
    on map, viii
    photographs of, plate 1 (top and
        bottom), plate 3 (bottom left)
    visitors to, 45, 52, 57–58
    wolves around, 63, 93–94, 101–2
Moscow, Idaho, 75
Mountain West, ix
Ms. Badass (wolverine), 82–83, 85
Munich Wildlife Society, 141
Murphy Lake Pack, 131

N
National Park Service, 113
Native Counselling Services of
    Alberta, 97
Naughty Pine Saloon, Montana, 184
Nettie Creek, British Columbia, 40
Nez Perce Tribe, 131, 171, 194
Niemeyer, Carter, 128–29
Ninemile drainage, Montana, 87
Ninemile Pack, 168
North American Wolf Symposium,
    168